Electronics Engineering

Electronics Engineering

Prof. O.N. Pandey
JSS Academy of Technical Education,
Noida

GEORGE GREEN LIBRARY OF
SCIENCE AND ENGINEERING

Taylor & Francis Group
Boca Raton London New York

CRC is an imprint of the Taylor & Francis Group,
an informa business

Ane Books Pvt. Ltd.

Electronics Engineering

O.N. Pandey

© Ane Books Pvt. Ltd.

First Published in 2010

Ane Books Pvt. Ltd.
4821 Parwana Bhawan, 1st Floor
24 Ansari Road, Darya Ganj, New Delhi -110 002, India
Tel: +91 (011) 2327 6843-44, 2324 6385
Fax: +91 (011) 2327 6863
e-mail: anebooks@vsnl.net
Website: www.anebooks.com

For

CRC Press
Taylor & Francis Group
6000 Broken Sound Parkway, NW, Suite 300
Boca Raton, FL 33487 U.S.A.
Tel : 561 998 2541
Fax : 561 997 7249 or 561 998 2559
Web : www.taylorandfrancis.com

For distribution in rest of the world other than the Indian sub-continent

ISBN : 978 1 43981 4413

All rights reserved. No part of this publication may be reproduced, stored in a retrieval system, or transmitted in any form or by means, electronic, mechanical, photocopying, recording and/or otherwise, without the prior written permission of the publishers. This book may not be lent, resold, hired out or otherwise disposed of by way of trade in any form, binding or cover other than that in which it is published, without the prior consent of the publishers.

British Library Cataloguing in Publication Data
A catalogue record for this book is available from the British Library

Printed at Thomson Press, Delhi

Dedicated to my Parents
Smt. Kalavati Devi
and
Shri Deo Murti Pandey
Who have given so much to me

Preface

Development in Electronics Engineering are taking place today at an awesome place. Therefore, it has become essential to understand the fundamentals of Electronics. This book presents fundamentals of electronics.

The book has been written after going through a large number of references. The objective has been to present the matter in simple, straight forward and easy form without losing any important information and detail. Appendices have been added to cover electronic symbols, abbreviations, diagrammatic symbols, various parameter units, conversion factors, periodic table of elements, conduction properties round copper conductor data, standard resistors and capacitors, electronic formulae, equivalent circuits and characteristics. Glossary of electronic terms has been added for quick understanding of electronic terms.

I am very thankful to Dr. Narendra Kumar and Prof. Dinesh Chandra of JSS Academy of Technical Education, Noida for encouragement leading to such technical contribution.

I appreciate the cooperation and help extended by my wife Mrs. Ranjana Pandey, Son Nishith Pandey, relatives and friends. I also appreciate the efforts and help extended by Mrs. Shilpi Gadi, Mrs. Amita Rana, Ms. Paro Bajpai, Mr. Rahul Gupta and others.

— **Prof. O.N. Pandey**

Contents

Preface   *vii*

1. BASICS OF ELECTRONICS   1–20

    1.1   Introduction   1

    1.2   Electronic Charge and Current   2

    1.3   Electronic Circuit Components   2

    1.4   Voltage and Current Relationships   5

    1.5   Work, Power and Energy   6

    1.6   SI Units   7

    1.7   Voltage and Current Sources   8

    1.8   Semiconductor Materials   9

    1.9   *P-N* Junction and Depletion Layer   15

    Summary   17

    Exercises   19

2. SEMICONDUCTOR DIODES   21–68

    2.1   Semiconductor Diode   21

    2.2   *V-I* Characteristics   22

    2.3   Ge, Si and Gaas Characteristics   24

    2.4   Ideal and Practical *V-I* Characteristics   24

    2.5   Diode Resistance   25

2.6 Diode Ratings 29
2.7 *P-N* Junction (Diode) as Rectifiers 30
2.8 Ripples Efficiency and Regulation 37
2.9 Efficiency of Full-wave Rectifier 38
2.10 Filters for Rectifiers 39
2.11 Clipping Circuits 44
2.12 Clamping Circuits 47
2.13 Voltage Multipliers 48
2.14 Zener Diodes 50
2.15 Temperature Coefficient 54
2.16 Zener Diode Application as Shunt Regulator 55
2.17 Diodes for Optoelectronics 56
2.18 Other Types of Diodes 58
Summary 62
Exercises 65

3. BIPOLAR JUNCTION TRANSISTOR (BJT) 69–146

3.1 Basic Construction 69
3.2 Transistor Action 70
3.3 Circuit Configurations 72
3.4 Input/Output Characteristics 74
3.5 Mathematical Relationships 78
3.6 Biasing of Transistors 79
3.7 Graphical Analysis of CE Amplifier 87
3.8 Parameter Model 90
3.9 Hybrid Equivalent Circuit for Common Base (CB) 95
3.10 Hybrid Equivalent Circuit for Common Collector (CC) 95

3.11	Overall Current Gain	96
3.12	Overall Voltage Gain	97
Summary		142
Exercises		144

4. FIELD EFFECT TRANSISTOR (FET) — 147–186

4.1	Introduction	147
4.2	Junction Field Effect Transistor (JFET)	148
4.3	Working Principle of JFET	149
4.4	Concept of Pinch-Off and Maximum Drain Saturation Current	151
4.5	Input and Transfer Characteristics	152
4.6	Parameters of JFET	153
4.7	JFET Biasing	154
4.8	JFET Connections	156
4.9	Metal Oxide Semiconductor Field Effect Transistor (MOSFET)	164
4.10	MOSFET Operation	166
4.11	Characteristics of MOSFET	168
Summary		182
Exercises		184

5. OPERATIONAL AMPLIFIER (OP-AMP) — 187–220

5.1	Introduction	187
5.2	Op-amp Integrated Circuit	187
5.3	Op-amp Symbol	188
5.4	Concept of Ideal Op-amp	189
5.5	Inverting Amplifier	189
5.6	Non-inverting Amplifier	190

5.7	Unity Gain or Voltage Follower Amplifier	191
5.8	Op-amp as Adder or Summer	192
5.9	Op-amp as Difference Amplifier	193
5.10	Subtractor	195
5.11	Differentiator	195
5.12	Integrator	196
5.13	Op-amp Parameters	197
Summary		217
Exercises		218

6. SWITCHING THEORY AND LOGIC DESIGN (STLD) — 221–268

6.1	Introduction	221
6.2	Number System	221
6.3	Conversion of Bases	224
6.4	Binary Coded Decimal (BCD) Numbers	227
6.5	Binary Addition	227
6.6	Binary Subtraction	227
6.7	Boolean Algebra	227
6.8	Boolean Algebra Theorems Table	229
6.9	Logic Gates and Universal Gates	229
6.10	Canonical Forms	231
6.11	K-map	232
6.12	Simplification of Boolean Expression using K-map	233
6.13	Simplification in Sum of Product (SOP) form	234
Summary		263
Exercises		265

7. ELECTRONICS INSTRUMENTS — 269–278

    7.1 Digital Voltmeters (DVMs)   269

    7.2 Digital Multimeters (DMMs)   272

    7.3 Cathode Ray Oscilloscope (CRO)   273

    7.4 Measurements Using CRO   275

    Summary   276

    Exercises   277

Appendix   **279**

Glossary   **307**

Index   **323**

1

BASICS OF ELECTRONICS

1.1 INTRODUCTION

Electron mechanics is known as Electronics. Electronics puts electrons to work using the science and technology of the electron motion. The advancement of electronics has been vary fast. Electronics has given tremendous growth in computer science, communication, control, instrumentation, information technology. Although electronic devices such as computer, cellular phone or television, *etc.*, are well-known, but inside of these devices are a mystery. Electronics engineering is the knowledge related with functioning of electronic devices.

Development of electronics started with vacuum diode in 1897 and vacuum triode in 1906. Semiconductor electronics started with the invention of transistor in 1948 and this replaced tube based electronics. The electronic components developed are diode, transistor, field effect transistor (FET).

Integrated circuits (ICs) were developed in 1958. ICs are basically an entire electronic circuit on a single semiconductor chip. A single chip has all active and passive components and their interconnections integrated during manufacturing process. ICs drastically reduced the size, weight and cost of the electronic devices.

The design and fabrication of high density ICs is known as microelectronics. The small scale integration (SSI) have components less than 100, medium scale integration (MSI) have 100 to 1000 components, large scale integration have 1,000 to 10,000 components and very large scale integration (VLSI) have more than 10,000 components. New IC concepts resulted in new computer architecture which is based on speed, power consumption and component density. Thus, digital integrated circuits came into existence leading to transistor-transistor logic (TTL), emitter-coupled logic ECL, *etc*. The latest electronic component fabrication uses complementary metal oxide semiconductor (CMOS) technology.

The memories based on electronics are random access memories (RAMs) which are capable of both storing and retrieving data. RAMs store about 100 bits of information. 1600-bit, 64000-bit and 288000 bit RAMs have been developed using metal-oxide semiconductor (MOS) technology. More than a billion-bit RAM chips are available now. Further, read-only memories (ROMs), programmable ROMs (PROMs), erasable PROMs (EPROMs) are also available. Microprocessor (MP) development led to the "computer on a chip". Other developments due to MOS technology are charge-coupled device (CCD) which are being used in camera manufacturing, image processing and communication. Analog integrated circuits developed are operational amplifier (Op-amp), digital-to-analog (D/A), analog-to-digital (A/D) converters, analog multiplexer and active filters.

Electronics Engineering developments are taking place today at an awesome pace, therefore, it has become essential to understand the fundamentals of electronics.

1.2 ELECTRONIC CHARGE AND CURRENT

The smallest particle of any material is a molecule and subdivision of molecules are atoms. An atom consists of electrons, protons and neutrons. Electrons have negative charge, protons have positive charge and neutrons have no charge at all. An atom is electrically neutral, as number of its electrons is equal to number of protons. Bonding together have some loosely bound electrons *i.e.*, free electrons, silver, copper, aluminium and zinc materials have free electrons, therefore, it is easy to make them move. Such materials are known as **Conductors**. There are materials like glass, mica and porcelain which have closely bound atoms and movement of electrons from atoms is very difficult. Such materials are non-metallic and are known as **Insulators**.

An **electric current** is the movement of electrons along a definite path in a conductor. It is defined as:

$$\text{Current,} \quad i(t) = \frac{dq}{dt}$$

where
$i(t)$ = instantaneous current in amperes.
q = electric charge in coulombs.
t = time in seconds.

1.3 ELECTRONIC CIRCUIT COMPONENTS

Electronic circuit components are of two types: active and passive. **Active components** are semiconductor devices such as diodes, transistors, SCRs, FETs, *etc*. **Passive components** are resistors, inductors and capacitors. The active components shall be discussed in subsequent chapters, but passive components need to be discussed here itself.

1.3.1 Resistors

Resistance is a property of a conductor which opposes the flow of an electric current and it is denoted by R.

$$R = \rho \frac{l}{a} \text{ ohm or } \Omega$$

where l = length of the conductor in meter
 a = area of cross-section in meter^2
 r = specific resistance or resistivity of the material in ohm-meter.

Conductance, $\quad G = \dfrac{1}{R}$ mho or ℧

Most common resistors are moulded-carbon composition type. These are available in wattage ratings of $\frac{1}{4}$ W, $\frac{1}{2}$ W and 1 W with values from few ohms to 22 $M\Omega$. It has 5 to 20% tolerance. There are some resistors which are known as metal film resistors which have accuracy of ± 1%. These are also known as precision type. All these resistors are of very small size wherein printing of the ratings not feasible. Hence, colour coding done is as per Fig. 1.1.

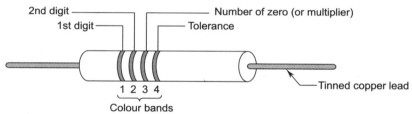

Fig. 1.1 Colour coding for resistor values.

The colour codes are given in three bands with fourth band for tolerance. The colour coding is given in the Table 1.1.

Table 1.1 Colour Coding of Resistors

S. No.	Colour	Digit	Multiplier	Tolerance	S. No.	Colour	Digit	Multiplier	Tolerance
1	Black	0	10^0	–	8	Violet	7	10^7	–
2	Brown	1	10^1	–	9	Grey	8	10^8	–
3	Red	2	10^2	–	10	White	9	10^9	–
4	Orange	3	10^3	–	11	Cold	–	–	± 5%
5	Yellow	4	10^4	–	12	Silver	–	–	± 10%
6	Green	5	10^5	–	13	No-colour	–	–	± 20%
7	Blue	6	10^6	–					

The above colour coding can be memorized as follows: all capital letters stand for colours.

> **B B ROY went to Great Britain and brought a Very Good Wife.**

Suppose 1st, 2nd, 3rd and 4th colours are yellow, violet, orange and silver respectively. What is the resistance value? The resistance value is given by:

Resistance value, $\quad R = 47 \times 10^3 \, \Omega \pm 10\%$

or $\quad\quad\quad\quad\quad\quad\quad\quad R = 4.7 \, k\Omega \pm 10\%$

The above were fixed resistors, but there are variable resistors in the form of Rheostats and Potentiometers. The resistors discussed so far have positive temperature coefficients *i.e.*, resistance value increases if the surrounding temperature increases. But, there are resistors which have negative temperature coefficients *i.e.*, resistance value decreases if the surrounding temperature increases. Such type of resistors are known as thermistors. These are made of semiconductor such as germanium (Ge) or silicon (Si). Other resistor types are Light Dependent Resistor (LDR) and Voltage Dependent Resistor (VDR). LDR resistance value depends on the intensity of light falling on it, therefore, it is also known as photoresistive cell or photoresistor. LDR is made of cadmium sulfide (CDS) or cadmium selenide (CdSe). VDR is based on junction field-effect transistor (JFET) which has three terminals, namely drain (D), source (S) and gate (G). The resistance between drain and source terminals is dependent on the gate voltage.

1.3.2 Inductors

Inductors store energy in the form of magnetic field. It has a winding of a conducting wire over a core which can be made of iron or just air itself. The current flowing through the coil establishes a magnetic field through the core. Inductor field reacts so as to oppose any change in current. The unit of inheritance is henry (H).

There are various types of inductors based on usage such as filter chokes which smoothens pulsating current produced by a rectifier. Audio-frequency chokes provide high impedance audio frequencies *i.e.*, between 60 Hz to 5 KHz. Variable inductors are used in turning circuits for radio frequencies.

1.3.3 Capacitors

A **capacitor** stores energy in the form of electric field. A capacitor consists of two conducting plates separated by an insulating material called **dielectric**. Capacitors can be of fixed value or variable value. The unit of capacitance is Farad (F). A capacitor opposes any change in the potential difference or voltage applied across its terminals.

There are Mica Capacitors which can be used upto 500 volts and are available in the range from 5 to 10,000 pF. There are ceramic capacitors which can be used in the range of 3 V to 6000 V. The capacitance value ranges from 3 pF to 3 µF. Such capacitors can be used in ac as well as dc circuits. Another type in paper capacitor which can be used from 100 V to several thousand volts. The capacitance values range from 0.0005 µF to several mF. Such capacitors can be used for both ac and dc circuits. Electrolytic capacitors are also available which can be used from 1 V to 500 V or more. The values may range from 1 μF to several thousand μF. These are marked with positive and negative terminals as such used mostly for dc circuits. Variable capacitors are also available wherein dielectric is air-gap and its variation leads to variation in the capacitance value.

1.4 VOLTAGE AND CURRENT RELATIONSHIPS

The relationships between potential difference across passive elements and the current through them is given here in the Table 1.2.

Table 1.2 Relationships between Voltages and Currents

S. No.	Passive Element	Relationship	Symbolic Circuits
1.	Resistor (R)	$v = iR$	
2.	Inductor	$v = L \dfrac{di}{dt}$	
3.	Capacitor	$i = C \dfrac{dv}{dt}$	

Resistor dissipates energy in the form of heat, inductor stores energy in the form of magnetic field and capacitor stores energy in the form of electric field. The voltage and current in the case of resistors are in phase, whereas in the case of inductors current lags voltage and in the case of capacitors current leads the voltage.

Heat produced by resistors, $H = I^2 Rt$ Joules

where
I = rms value of current in amperes (A)
R = resistance in ohms (Ω)
t = time in seconds (s)

Energy stored in Inductor, $E = \dfrac{1}{2} LI^2$ Joules

where L = inductance in Farads (F)

Energy stored in capacitor, $E = \dfrac{1}{2} CV^2$ Joules.

where
V = voltage across the capacitor in volts (V)
C = capacitance in Farads (F).

Conversion of Joule and Calorie:
1 calorie = 4.18 Joules.

1.5 WORK, POWER AND ENERGY

Workdone = Force × distance

or $\quad W = F \times d$ Joules

where
W = work done in Joules (J)
F = force applied in Newtons (N)
d = distance moved in meters (M)

Power = Rate of work done in Joules/second or watts (W).

or $\quad P = \dfrac{dW}{dt}$

$W = VQ$ Joules

where
V = voltage in volts (V)
Q = electric charge in columbs (C)

$$\text{Power} = \frac{\text{Energy}}{\text{time}} = \frac{W}{t} = \frac{VQ}{t} \text{ watts.}$$

Current, $\quad I = \dfrac{Q}{t}$ Amperes.

∴ Power, $\quad P = VI = I^2R = \dfrac{V^2}{R}$ watts.

Important conversions are:

1 hp (British) = 746 watts
1 hp (Metric) = 735.5 watts

Kinetic energy = $\dfrac{1}{2} mv^2$

where
m = mass of the material
v = velocity of the mass

Gravitational potential energy = mgh

where
m = mass of the material
g = gravitational acceleration, i.e., 9.81 m/s²
h = height by which mass is lifted.

Electric energy = power × time

or $\quad W = VI\,t = I^2R\,t = \dfrac{V^2}{R} t$

Electrical energy conversions are:

$$1 \text{ unit} = 1 \text{ kWh} = \frac{\text{Watts} \times \text{hour}}{1000}$$

$$1 \text{ kWh} = 3.6 \times 10^6 \text{ Joules} = 3.6 \text{ MJ}.$$

1.6 SI UNITS

The **SI units** are as per international system of units which are commonly used. The basic SI units are given in Table 1.3.

Table 1.3 Basic SI units

S. No.	Parameter	SI unit	Symbol
1.	Length	meter	m
2.	Mass	kilogram	kg
3.	Time	second	s
4.	Electric current	ampere	A
5.	Absolute temperature	Kelvin	K
6.	Luminous intensity	candela	Cd
7.	Amount of substance	mole	mol.

Temperature in Kelvin = 273 + temperature °C and the unit change in both units are 1 K and 1°C respectively.

Complete revolution = 2π radians or 360°

∴ 2π radians = 360°.

The various prefixes used in units are given in Table 1.4.

Table 1.4 Prefixes used in units

S. No.	Prefix	Multiplication factor	Symbol
1.	pico	10^{-12}	p
2.	nano	10^{-9}	n
3.	micro	10^{-6}	m
4.	milli	10^{-3}	m
5.	kilo	10^{3}	k
6.	mega	10^{6}	M
7.	giga	10^{9}	G
8.	tera	10^{12}	T

Some derived S.I. units are given in Table 1.5.

Table 1.5 Derived SI units

S. No.	Parameter	S.I. Unit	Symbol
1.	Area	Square meter	m^2
2.	Volume	Cubic meter	m^3
3.	Linear velocity	meter per second	m/s
4.	Angular velocity	radian per second	rad/s
5.	Linear acceleration	meter/second square	m/s^2
6.	Angular acceleration	radian/second square	rad/s^2
7.	Force	kilogramme meter per second square or or Newton	$kg\ m/s^2$ or N
8.	Weight = mass × gravitational acceleration $g = 9.81\ m/s^2$	kilogramme force or 9.81 Newtons	kgf or 9.81 N

1.7 VOLTAGE AND CURRENT SOURCES

Voltage sources are power supplies such as batteries, alternators, dynamos, etc. Metadyne generators, photoelectric cells, collector circuits of transistors are **current sources**. All these are known as independent voltage and current sources respectively. The **ideal voltage and current sources** are shown in Fig. 1.2. Ideal voltage source has zero internal resistance in series whereas ideal current source has infinite resistance in parallel with the current source.

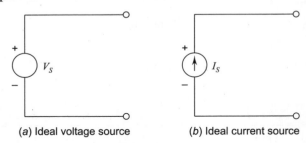

(a) Ideal voltage source   (b) Ideal current source

Fig. 1.2 Ideal Voltage and Current source.

The realistic voltage source has a internal resistance in series and realistic current source has a resistance in parallel as shown in Fig. 1.3.

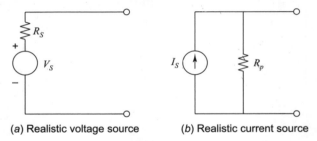

(a) Realistic voltage source   (b) Realistic current source

Fig. 1.3 Realistic Voltage and Current source.

Basics of Electronics

Conversion of voltage source to current source is shown in Fig. 1.4. It can be observed that:

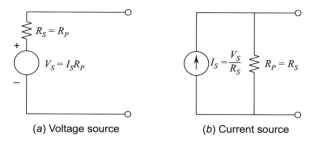

(a) Voltage source   (b) Current source

Fig. 1.4 Conversion between voltage and current source.

When voltage source is converted into current source, then current source values are:

$$I_S = \frac{V_S}{R_S} \quad \text{and} \quad R_P = R_S$$

when current source is converted into voltage source, then voltage source values are:

$$V_S = I_S R_P \quad \text{and} \quad R_S = R_P$$

when voltage or current source values are dependent on voltage or current values of a branch, then the voltage or current source are known as dependent voltage and current source as shown in Fig. 1.5.

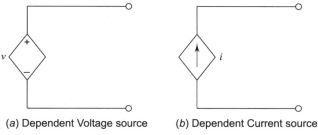

(a) Dependent Voltage source   (b) Dependent Current source

Fig. 1.5 Dependent voltage and current source.

1.8 SEMICONDUCTOR MATERIALS

A material is made up of one or more elements and an element is a substance composed entirely of atoms. The atoms of different elements differ in their structures, therefore, different elements have different characteristics.

An atom is comprised of a relatively massive core or nucleus carrying a positive charge, around which electrons move in orbits at distances which are great compared with the size of the nucleus. The electron mass is 9.11×10^{-31} kg and electron charge is -1.602×10^{-19} Coulomb $= -e$. The nucleus of every atom except that of hydrogen consists of protons and neutrons. Each proton carries a positive charge, e equal in magnitude to that of an electron and its mass is

1.673×10^{-27} kg *i.e.*, 1836 times that of electron. A neutron has no charge and its mass is almost same as that of a proton.

Hydrogen atom has the simplest structure as shown in Fig. 1.6. It consists of only a nucleus of one proton and one electron which revolves in an orbit, of 10^{-10} m diameter around proton. Atomic number of hydrogen is 1. Fig. 1.7 shows atomic structure of silicon (Si). Silicon's atomic number is 14. The electrons are arranged in orbits or shells. First orbit can have maximum of two electrons. The second orbit can have maximum of eight electrons. The third orbit can have maximum of eighteen electrons. The fourth orbit can have maximum thirty-two electrons. The uppermost orbit in an atom cannot have more than eight electrons.

Fig. 1.6 Hydrogen atom.

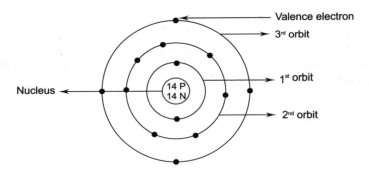

Fig. 1.7 Silicon atom.

The number of electrons present in the uppermost orbit is known as **valence electrons**. Silicon has four valence electrons. Fig. 1.8 shows a germanium (Ge) atom structure which has four valence electrons.

Atoms that have four valence electrons are known as tetravalent, and those with three are known as trivalent. Atoms with five valence electrons are known as pentavalent. The term valance indicates that the ionization potential required to remove any one of these electrons from atomic structure is significantly lower than that required for any other electron in the structure.

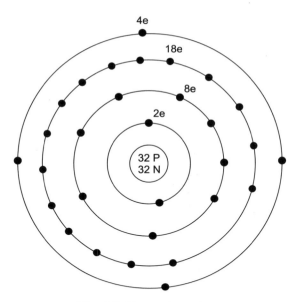

Fig. 1.8 Germanium atom.

Energy Levels

Within the atomic structure of each and every isolated atom, there are specific **energy levels** associated with each shell and orbiting electron as shown in Fig. 1.9. The farther an electron is from the nucleus, the higher is the energy state, and any electron that has left its parent atom has a higher energy state than any electron in the atomic structure. Fig. 1.9(a) shows that in an isolated atom discrete energy levels can exist. Fig. 1.9(b) shows a conductor wherein energy levels overlap, hence, electrons are free to move. Fig. 1.9(c) shows an insulator wherein the electrons require very high energy to bring them to conduction level. Fig. 1.9(d) shows insulators wherein a minimum energy level is associated with electrons in the conduction band and a maximum energy level of electrons bound to the valence shell of the atom. Between the two is an energy that the electron in valence band must overcome to become free carrier. This energy gap is different for Ge, Si, and GaAs. Ge has the smallest gap and GaAs the largest gap. In short, an electron in the valence band of silicon must absorb more energy than one in the valence band of germanium to become a free carrier. Similarly, an electron in the valence band of gallium arsenide must gain more energy than one in silicon or germanium to enter the conduction band.

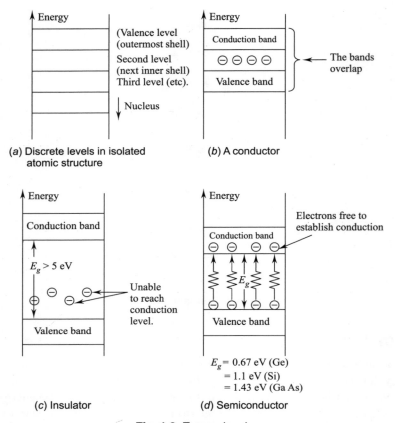

Fig. 1.9 Energy levels.

Thus, we can see that semiconductors are a special class of elements having a conductivity between that of a good conductor and that of an insulator. A semiconductor is an element with a valence of four *i.e.*, an isolated atom of the material has four electrons in its outer or valance orbit. The number of electrons in the valence orbit is the key to electrical conductivity. Conductors have one valence electron and insulators have eight valence electrons. The valance electrons get themselves detached from the nucleus on the application of small electric field. These free electrons constituting the flow of current are called conduction electrons.

There are two types of semiconductors. One is pure type, known as intrinsic type and other is impure type, known as extrinsic type. The conductivity of intrinsic semiconductor is poor at room temperature. Therefore, it is not used in electronic devices. Intrinsic semiconductors properties can be varied by adding impurities and their conduction properly can be varied by varying temperature.

1.8.1 Intrinsic Semiconductors

In **intrinsic semiconductors** even at room temperature, some of the valance electrons may acquire enough energy to cross over to conduction band from valence band, thereby becoming free electrons. As the electrons leave the valance band it creates a vacant space in it. This is known as **'hole'**. Thermal energy produces free electrons and holes in pairs. In Fig. 1.10, a dc voltage is applied which will force the free electrons to move left and the holes to flow right. When the free electrons arrive at left end of the semiconductor crystal, they enter the external wire and flow to the positive battery terminal. On the other hand, the free electrons at the negative battery terminal will flow to the right end of the crystal. At this point, they enter the crystal and recombine with holes that arrive at the right end of the crystal. In this way, a steady flow of free electrons and holes occur inside the semiconductor. The current in a semiconductor is the combined effect of the flow of free electrons in one direction and the flow of holes in the other direction. Free electrons and holes are called carriers as they carry a charge from one place to another.

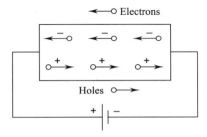

Fig. 1.10 Intrinsic semiconductors.

1.8.2 Extrinsic Semiconductors

Extrinsic semiconductors are created by a process of adding impurities deliberately to an intrinsic semiconductor. The process is known as doping. The added impurity is known as doping agent. When doped with a trivalent impurity, the impurity accepts one electron to achieve stable state. This type of doping agent is known as an **acceptor**. When doped with a pentavalent impurity, the impurity donates one electron to the conduction band. This type of doping agent is known as **donor**.

There are three semiconductors most frequently used in the construction of electronic devices namely Si (silicon), Ge (germanium) and GaAs (gallium arsenide) Si and Ge are single-crystals whereas GaAs is a compound crystal. Initially, Ge was easily available and refinement, *i.e.*, process of purity was easy. Therefore, Ge was used for first few decades. However, it was found that Ge was very sensitive to changes in temperature. In 1954, the trial of Se was done which was less sensitive to temperature and available in abundance. Therefore, Si became most popular choice. When speed of operation and communication on computers became the basic requirement, GaAs became very handy in 1970s. Thus, GaAs is used as the basic material for new high-speed and very large scale integrated circuits (VLSI).

1.8.2.1 N-type semiconductor

A pentavalent (phosphorous) impurity addition to an intrinsic semiconductor (silicon) gives N-type semiconductor. Phosphorus has five valance electrons and silicon has four valance therefore, in this case one free electron is available as shown in Fig. 1.11.

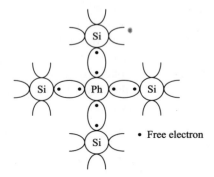

Fig. 1.11 N-type semiconductor *i.e.*, one free electron without a hole.

Thus, every phosphorus atom contributes one free electron without creating a hole. Consequently, number of free electrons becomes far greater than the number of holes. Such extrinsic semiconductor is known as N-type semiconductor.

1.8.2.2 P-type of semiconductor

A trivalent (boron, aluminium, *etc.*) impurity addition to an intrinsic semiconductor (silicon), gives P-type semiconductor.

Boron has three valance electrons and silicon has four valance electrons, therefore, the deficiency of an electron around the boron atom gives rise to a hole, see Fig. 1.12. Thus, every boron atom contributes one hole. Hence, number of holes become far greater than the number of electrons. This results in P-type semiconductor.

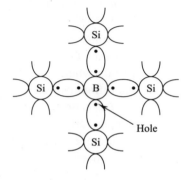

Fig. 1.12 P-type semiconductor *i.e.*, one hole is available per atom of boron.

1.9 P-N JUNCTION AND DEPLETION LAYER

1.9.1 P-N Junction

P-N junction is formed by a special fabrication technology. To make a *P-N* junction, the *N*-type and *P*-type semiconductor crystals are cut into thin slices called wafers.

If a wafer of *P*-type semiconductor is joined to a wafer of *N*-type semiconductor in such a manner that the crystal structure remains continuous at the boundary, then a new structure called *P-N* junction is formed.

1.9.2 Depletion Layer

In a *P-N* junction, the *P*-region has holes and negatively charged impurity ions. *N*-region has free electrons and positively charged impurity ions. Electrons and holes are mobile charges whereas the ions are immobile. When a *P-N* junction is formed, the holes in the *P*-region diffuse into *N*-region and the electrons in the *N*-region diffuse into *P*-region. This process is called diffusion which happens for a short time as soon as the *P-N* junction is formed. After a few combinations of holes and electrons, a restraining force is developed which is known as **potential barrier**. This potential barrier prevents further diffusion of holes and electrons. The barrier force development can be easily explained. That is, each recombination of hole and electron eliminates hole and electron. During this process the negative acceptor ions in the *P*-region and positive donor ions in the *N*-region are left uncompensated. The additional holes trying to diffuse into *N*-region and additional electrons trying to diffuse into *P*-region are repelled by these negative and positive charges respectively. The region containing this uncompensated acceptor and donor ions is called **depletion layer**, see Fig. 1.13.

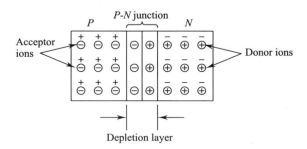

Fig. 1.13 Formation of *P-N* junction and depletion layer.

1.9.3 Forward Biasing

When an external voltage is applied to the *P-N* junction in such a direction that it cancels the potential barrier which permits current flow, it is called **forward biasing**.

Fig. 1.14 shows forward biasing connections. Positive terminal of battery is connected to *P*-type and negative terminal to *N*-type. The applied forward potential establishes an electric field which acts against the field due to depletion (potential) barrier. Thus, the depletion (potential) barrier is reduced and allows the flow of charged carriers across the barrier. In effect, a small forward voltage is sufficient to make the depletion barrier insignificant. Once the depletion barrier is made insignificant by the forward voltage, junction resistance becomes too small and a high current flows in the circuit.

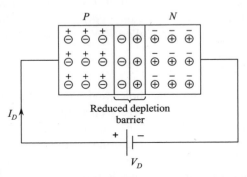

Fig. 1.14 Forward biasing reduces, depletion (potential barrier).

1.9.4 Reverse Biasing

When the external voltage applied to the *P-N* junction is in such a direction that depletion (potential) barrier is increased, it is called **reverse biasing**.

Fig. 1.15 shows reverse biasing connection. Negative terminal of battery is connected to *P*-type and positive terminal to *N*-type. The reverse biasing establishes an electric field which acts in the same direction as the field due to depletion (potential) barrier. Thus, depletion (potential) barrier is increased and prevents the flow of charge carriers across the junction. In effect, a high resistance path is established for the circuit, hence, the current flow is insignificant.

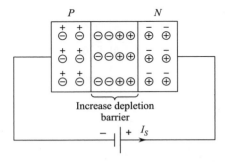

Fig. 1.15 Reverse bias increases depletion barrier (potential barrier).

Basics of Electronics 17

SUMMARY

1. **Semiconductor Devices:** Semiconductor devices are diode, transistor and integrated circuits (ICs), ICs are known as Microelectronics. ICs can be small scale integration (SI), medium scale integration (MSI), large scale integration (LSI) and very large scale integration (VLSI). There are digital as well as analog ICs. Digital logic can be based on transistor-transistor logic (TTL), emitter-coupled logic (ECL), *etc.* Latest electronic components use complementary metal oxide semiconductor (CMOS) technology. The semiconductor memories are random access memory (RAM), read only memory (ROM), programmable ROM (PROM) and erasable PROM (EPROM). Microprocessor (MP) has given bite to "Computer on Chip". Operational amplifier (op-amp) is an analog IC. Other electronic devices are digital to analog (D/A) converter, analog-to-digital (A/D) converter, analog multiplexers and active filters.

2. **Electronic Circuit Components:** There are two types of electronic circuit components. One type of components are active components such as diodes, transistors, silicon controlled rectifiers (SCRs), field-effect-transistors (FETs), etc. Other type of components are passive components such as resistors inductors and capacitors.

3. **SI Units:** SI units are as per international system of units. The basic SI units are meter, kilogram, second, ampere, kelvin, candela and mole.

    Temperature in Kelvin = 273 + temperature in °C.

4. **Voltage and Current Sources:** Voltage source has very small internal resistance and ideal voltage source has zero internal resistance. Current source has very large resistance across it and ideal current has infinite resistance across it. Voltage and current sources can be converted from one to other or *vice-versa* for circuit analysis purposes.

5. **Semiconductor Materials:** Number of electrons in uppermost orbit of an atom of a material, is known as valence. Conductors have one valance electron, insulators have eight valence electrons and semiconductors have four valence electrons. Each silicon atom in a crystal has its four valence electrons plus four more electrons that are shared by the neighbouring atoms.

6. **Intrinsic Semiconductors:** It is a pure semiconductor. When an external voltage is applied to the intrinsic semiconductor, the free electrons flow towards the positive battery terminal and the holes flow towards the negative battery terminal.

7. **Extrinsic Semiconductors:** When an intrinsic semiconductor is doped with pentavalent donor atoms, it has more free electrons than holes. When an intrinsic semiconductor is doped with trivalent acceptor atoms, it has more holes than free electrons.

8. **N-type and P-type semiconductors:** In an N-type semiconductor the free electrons are the majority carriers, while the holes are the minority carriers. In a P-type semiconductor the holes are the majority carriers, while the free electrons are the minority carriers.

9. **Forward and Reverse Biasing:** When a battery is connected across the P-N junction, then this process is known as biasing of the P-N junction. In forward biasing of a P-N junction, positive terminal of the battery is connected to the P-side and the negative terminal to the N-side. In reverse biasing the positive terminal of battery is connected to N-side and negative side is connected to P-side of the P-N junction.

10. **Important Formulae:**

    (i) $v = iR$

    (ii) $v = L\dfrac{di}{dt}$

    (iii) $i = C\dfrac{di}{dt}$

    (iv) Heat produced by resistor $= I^2 R\, t$ Joules

    (v) Energy stored in an inductor $= \dfrac{1}{2} LI^2$ Joules.

    (vi) Energy stored in a capacitor $= \dfrac{1}{2} CV^2$ Joules.

    (vii) 1 calorie = 4.18 Joules.

    (viii) Power $= VI = I^2 R = \dfrac{V^2}{R}$ watts.

    (ix) 1 hp (British) = 746 watts.

    (x) 1 hp (Metric) = 735.5 watts

    (xi) 1 unit energy = 1 kWh = 3.6 MJ.

EXERCISES

1.1 What do you understand by electronics ? Explain its utility in our daily life.

1.2 Explain latest trends in electronics.

1.3 What do you understand by electric current ?

1.4 What are active components ? Name three active components.

1.5 What is a resistor ? What is the relationship of resistance value with length and of cross section of a conductor ?

1.6 Explain colour coding of a resistor with an example.

1.7 Explain inductors and capacitors. What are their relationships with current through and voltage across there ?

1.8 Explain electrical power, energy and their relationships with current and voltages.

1.9 What is SI unit? Explain basic and derived SI units.

1.10 Write short note on voltage and current sources.

1.11 Explain an atom with a diagram.

1.12 What do you understand by valence of a material ? Give valences of conductor, insulator and semiconductor.

1.13 What is an intrinsic semiconductor ?

1.14 Explain extrinsic semiconductor.

1.15 What are *N*-type and *P*-type semiconductors ?

1.16 Explain a *P-N* junction and depletion layer.

1.17 What do you understand by forward and reverse biasings ?

2

SEMICONDUCTOR DIODES

2.1 SEMICONDUCTOR DIODE

A *P-N* junction is known as **semiconductor diode**. A semiconductor diode is represented by the schematic symbol shown in Fig. 2.1. The arrow indicates the direction of forward bias current flow. It has two terminals.

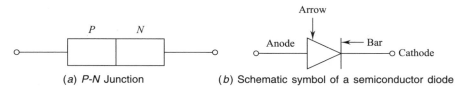

(a) P-N Junction (b) Schematic symbol of a semiconductor diode

Fig. 2.1 *P-N* junction and schematic symbol of a diode.

If the *dc* power supply pushes current in the direction of arrow, it is **forward biased**; and if the current is trying to flow opposite to arrow direction, it is **reverse biased**. Fig. 2.2 shows forward and reverse biased circuits.

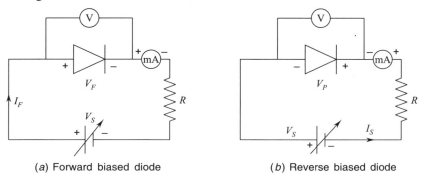

(a) Forward biased diode (b) Reverse biased diode

Fig. 2.2 Forward and reverse biased diodes.

The general characteristics of a semiconductor diode can be demonstrated through the use of solid-state physics. Shockley's equation represents these characteristics. For the forward and reverse bias regions, the equation is:

$$I_F = I_S\left(e^{V_F/\eta V_T} - 1\right)$$

where, V_F = applied forward bias voltage across the diode
I_F = forward-bias current through the diode
I_S = reverse bias saturation current
η = a factor representing operating conditions. For germanium diode $\eta = 1$ and for silicon diode $\eta = 2$
V_T = thermal voltage determined by $V_T = \dfrac{kT}{q}$

where, k = Boltzmann's constant = 1.38×10^{-23} J/Kelvin
T = absolute temperature in kelvins
= 273 + temperature in °C.
q = magnitude of electronic charge
= 1.6×10^{-19} Coulomb

At room temperature ($T = 300°K$), $V_T = 26$ mV

Thus, at room temperature, $I_F = I_S\left(e^{40V_F/\eta} - 1\right)$.

2.2 V-I CHARACTERISTICS

From equation (2.1), for forward bias V_F will be positive and current equation is given by:

$$I_F = I_S\, e^{V_F/\eta V_T} - I_S$$

The first term in the equation is very high as compared to I_S, hence, the forward current is;

$$I_F \cong I_S\, e^{V_F/\eta V_T}$$

The forward bias current for theoretical case is shown in Fig. 2.3 by dotted lines for a silicon diode

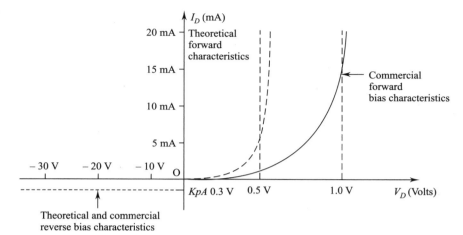

Fig. 2.3 V-I characteristics of ideal and practical diode.

For a voltage $V_F = 0$, the equation for current
$$I_F = I_S(e^0 - 1) = 0.$$
For a negative V_F (reverse-bias), the equation for current
$$I_F = I_S e^{-V_F/\eta V_T} - I_S$$
The first term will be too small as compared to I_S, hence, current becomes
$$I_F \cong -I_S$$
The theoretical characteristics are again shown in Fig. 2.3 by dotted line for $V_D = 0$ and reverse-bias. However, commercially available diode forward-bias diode characteristics differ from the theoretical due to internal body resistance and external contact resistance of diode, *etc*. Thus, commercial diode forward-bias diode characteristics are shown by continuous line in Fig. 2.3. The theoretical and commercial diode current in the reverse bias case are too small, *i.e.*, 10 pA to 1 µA, therefore, characteristics for negative V_D (reverse-bias) is almost same.

Example 2.1 The reverse saturation current at room temperature is 0.4 µA when a reverse bias is applied to a Ge diode. What is value of current flowing in the diode, if 0.15 V forward bias is applied at room temperature?

Solution: Given: $I_S = 0.4$ µA

Forward bias voltage, $V_F = 0.15$ V

∴ The current flowing through the diode under forward bias at room temperature:
$$I_F = I_S\left(e^{40V_F/\eta} - 1\right)$$

$h = 1$ for germanium diode

∴ $I_F = 0.4 \times 10^{-6}(e^{40 \times 015} - 1)$

or $I_F = 160.87$ µA.

2.3 GE, SI AND GAAS CHARACTERISTICS

The V-I characteristics considered so far have been for silicon diodes. The V-I characteristics for the three materials are shown in Fig. 2.4. The centre of the knee of the curve, *i.e.*, barrier potential is about 0.3 V for **Ge**, 0.7 V for **Si** and 1.2 V

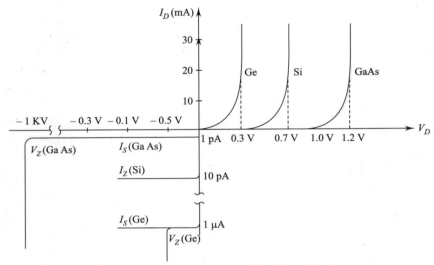

Fig. 2.4 V-I characteristics of Ge, Si and GaAs.

for **GaAs**. It can be seen that best characteristics are for GaAs and next good one is for Si; and the Ge is the last one, *i.e.*, least desirable. It is important to note in the reverse-bias case, there is a voltage V_z at which reverse bias current, suddenly jumps to very high current. V_z is known as zero potential. The zero voltages for Ge, Si and GaAs are – 50 V, – 100 V and – 1 KV respectively.

2.4 IDEAL AND PRACTICAL V-I CHARACTERISTICS

An ideal diode characteristics should be such that it allows full current to flow in forward-bias condition and zero current in reverse-bias condition. In other words, an ideal diode will act as a closed switch in forward-bias condition, whereas as an open switch in reverse-bias condition. Figure 2.5 shows an ideal semiconductor diode in forward-bias and reverse-bias condition.

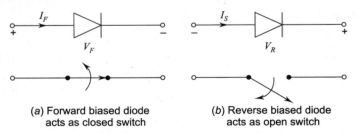

(a) Forward biased diode acts as closed switch

(b) Reverse biased diode acts as open switch

Fig. 2.5 An ideal semiconductor diode.

Semiconductor Diodes

The **ideal semiconductor V-I characteristics are** shown in Fig. 2.6. The semiconductor diode has zero resistance in forward-bias and infinite resistance in reverse-bias condition. **The actual V-I characteristics** will be as explained earlier for theoretical or commercial cases.

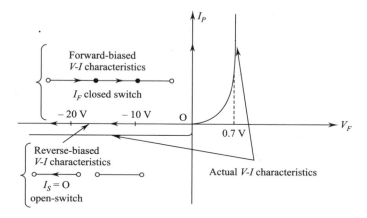

Fig. 2.6 An ideal and actual V-I characteristics of a semiconductor diode.

2.5 DIODE RESISTANCE

It is clear by now that a forward biased diode conducts easily whereas reverse biased diode conduction is negligible. In other words, forward resistance of a diode is very small as compared to its reverse resistance.

2.5.1 Forward Resistance

Forward biased diode resistance changes with the changing current, thus it can be *dc* forward resistance or *ac* forward resistance.

2.5.1.1 *DC forward resistance or static resistance (R_F)*

In the case of application where direct current flows, the forward diode resistance can be explained by Fig. 2.7. Suppose voltage applied is *OA* and *dc* current *OB* is flowing through the diode, then

DC **forward resistance,** $\boxed{R_F = \dfrac{OA}{OB}}$

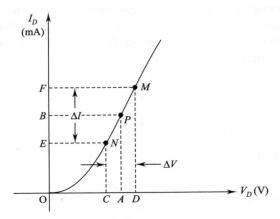

Fig. 2.7 DC forward bias resistance

2.5.1.2 AC forward resistance (r_F) or dynamic resistance

AC forward resistance is the resistance offered by the diode due to the changing current. Consider Fig. 2.7

AC **forward resistance,** $\quad \boxed{r_F = \dfrac{\text{Change in voltage across the diode}}{\text{Corresponding change in current}}}$

or $$r_F = \frac{OD - OC}{OF - OE} = \frac{CD}{EF} = \frac{\Delta V_F}{\Delta I_F}$$

AC forward resistance is significant as diodes are generally used with *ac* voltages. AC forward resistance of diode is very small in the range of 1–25 Ω.

The diode current equation is given by:

$$I_F = I_s\left(e^{V_F/\eta V_T} - 1\right)$$

By differentiating, we get $\quad \dfrac{dI_F}{dV_F} = I_S \times \dfrac{V_F}{\eta V_T} \times e^{V_F/\eta V_T}$

or $$\frac{dV_F}{dI_F} = \frac{\eta V_T}{I_S\, e^{V_F/\eta V_T}} = \frac{\eta V_T}{I_F + I_S}$$

∴ $$r_F = \frac{\eta V_T}{I_F + I_S}$$

or **diode resistance,** $\quad \boxed{r_F = \dfrac{\eta V_T}{I_F} \text{ as } I_S \ll I_F}$

Thus, for forward-bias, r_F is inversely proportional to I_F. At room temperature, *i.e.*, 27° (300°K), we have:

$$r_F = \frac{1 \times 26 \text{ mV}}{I_F \cdot \text{mA}} \text{ taking } \eta = 1 \text{ (germanium and } V_T = 26 \text{ mV)}$$

Example 2.2 Determine the dynamic resistance of a P-N junction diode at forward current of 2 mA. Assume that $\frac{kT}{e} = 2.5$ mV.

Solution:

Given: Forward current, $I_F = 2$ mA

Voltage equivalent of temperature,

$$V_T = \frac{kT}{e} = 2.5 \text{ mV}$$

We know that:

Dynamic resistance, $$r_F = \frac{\eta V_T}{I_F}$$

$$r_F = \frac{1 \times 2.5 \text{ mV}}{2 \text{ mA}} \text{ taking } \eta = 1$$

or $$r_F = 1.25 \ \Omega.$$

2.5.1.3 Reverse resistance (R_R)

The resistance of diode due to reverse bias is known as reverse resistance. The reverse resistance is too high, nearly infinite.

Reverse resistance, $R_R \simeq 40{,}000 \ R_F$ for germanium.

2.5.2 Transition and Diffusion Capacitance

An electronic circuit is sensitive to frequency. At high frequencies for the diode, stray capacitive effects are considerable. Forward-bias condition has the effect of diffusion leading to **diffusion capacitance** (C_D). In reverse-bias condition there is depletion-region or **transition capacitance** (C_T). The capacitive effects are represented by capacitors in parallel with the ideal diode, as shown in the Fig. 2.8. These are applicable normally in power areas.

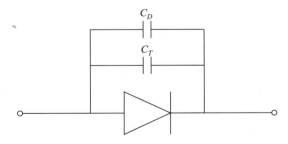

Fig. 2.8 Effect of capacitance on the semiconductor diode.

The diffusion capacitance is given by the formula:

$$C_D = \frac{I_F}{\eta V_T}$$

where,
η = constant (η = 1 for Ge and η = 2 for Si)
V_T = volt equivalent of temperature
T = mean life time of current
I_F = forward current

The depletion or transition capacitance is given by formula:

$$C_T = \frac{K}{\sqrt{V}} \quad \text{where} \quad V = \text{applied bias voltage.}$$

The capacitance V_S applied bias voltage plot is given below:

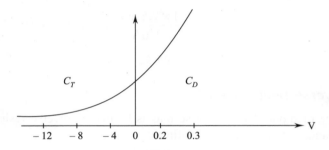

2.5.3 Diode Equivalent Circuit

The diode linear characteristic forward bias gives:

$$V_F = V_T + I_F r_F$$

The actual and linear characteristics along with **equivalent circuit** are as follows:

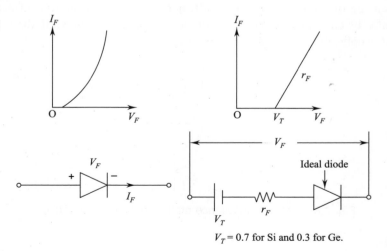

V_T = 0.7 for Si and 0.3 for Ge.

2.6 DIODE RATINGS

Data sheets of diode specify several useful parameters and some of these are explained here.

2.6.1 Repetitive Peak Current (I_{peak})

It is the maximum instantaneous value of repetitive forward-bias current.

2.6.2 Average Current (I_{av})

It is an average forward-bias current value and is defined by $I_{av} = 0.318\ I_{peak}$.

2.6.3 Peak-Inverse Voltage (V_R)

It is the absolute peak voltage which must be applied in reverse bias across the diode.

2.6.4 Steady-State Forward Current (I_F)

It is the maximum current which can be passed continuously through the diode.

2.6.5 Peak-Forward Surge Current (I_{FS})

It is a current which may flow briefly when a circuit in switch is first switched on. I_{FS} is very much higher than I_F.

2.6.6 Static Maximum Voltage Drop (V_{FM})

It is a maximum forward voltage drop for a forward current at the device temperature.

2.6.7 Continuous Power Dissipation (P)

It is the maximum power which can be dissipated continuously in free air.

2.6.8 Reverse Recovery Time (t_{rr})

It is the maximum time for the device to switch from ON to OFF.

Example 2.3 What is the current in the circuit shown below:

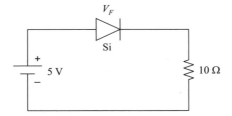

Solution: For silicon diode, $V_F = 0.7$ V

Using KVL in the circuit, we get:

$$5 = V_F + I \times 10$$

or

$$I = \frac{5 - 0.7}{10} = \frac{4.3}{10} = 0.43 \text{ A}$$

Thus, current in the circuit, $I = 0.43$ A.

Example 2.4 Find the voltage V_A and the current in the circuit shown in figure given below:

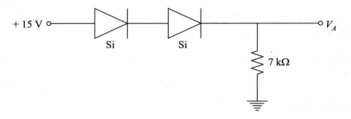

Solution: For silicon diode, $V_F = 0.7$ V

Voltage, $V_A = 15 - (0.7 \times 2) = 13.6$ V

and current, $I = \dfrac{13.6}{7 \times 10^3} = 1.942 \times 10^{-3}$ A   or   1.942 mA

2.7 P-N JUNCTION (DIODE) AS RECTIFIERS

The electrical power supply to Indian homes and industries is in the form of *ac* voltage. It is 220 volt *rms* at 50 Hz for domestic usage. The electronic equipments are operated by *dc* supply. It can be dry cells or **battery eliminator**. A battery eliminator gets *ac* voltage as input and converts it into *dc* supply. Battery eliminator is also known as *dc* power supply. The individual units in a *dc* power supply are input step down transformer, **rectifier**, filter and regulator. Figure 2.9 shows the block diagram of a *dc* power supply.

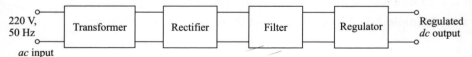

Fig. 2.9 Block diagram of a dc supply.

2.7.1 Half-Wave Rectifier

A half-wave rectifier is shown in Fig. 2.10. The *ac* supply is input to a step-down transformer. The secondary has a diode which is connected across the load as shown. Suppose secondary voltage is given as:

Semiconductor Diodes

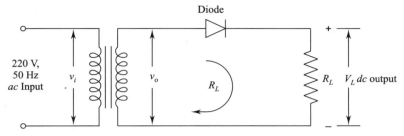

Fig. 2.10 Half-wave rectifier.

$$V_o = V_m \sin \omega t$$

Then half-wave rectifier wave-forms will be as shown in Fig. 2.11.

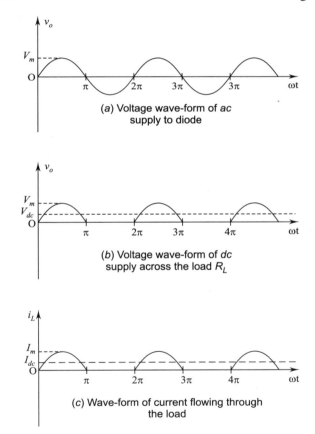

Fig. 2.11 Half-wave rectifier wave-forms.

The ac voltage across the secondary winding of the transformer changes polarity after every half cycle. During positive half cycle, the diode is forward biased, hence current flows through the load. In other words, during positive half-cycle voltage across the load is also positive half cycle. During negative half cycle, the diode is subjected to reverse-bias, hence, negligible current flows through the load *i.e.*, there is no voltage across the load. Thus, output across the load is pulsating *dc*.

The current through the load is given by:
$$i_L = I_m \sin \omega t \text{ for } 0 \le \omega t \le \pi$$
$$= 0 \text{ for } \pi \le \omega t \le 2\pi$$

Peak value of current, $I_m = \dfrac{V_L}{R_L}$

The average value of load current,
$$I_{av} = I_{dc} = \dfrac{\text{Area of wave for a cycle}}{\text{Duration of a cycle}} = \dfrac{\text{area}}{\text{base}}$$

We know that:
$$\text{Area} = \int_0^{2\pi} i_L \, d(\omega t)$$
$$= \int_0^{\pi} I_m \sin \omega t \, d(\omega t) + \int_{\pi}^{2\pi} 0 \, d(\omega t)$$
$$= I_m \left[-\cos \omega t\right]_0^{\pi} + 0$$
$$= I_m[-\cos \pi - (-\cos 0)]$$
$$= I_m [1 + 1] = 2 I_m$$

$\therefore \qquad I_{dc} = \dfrac{\text{area}}{\text{base}} = \dfrac{2 I_m}{2\pi}$

or $\qquad I_{dc} = \dfrac{I_m}{\pi}.$

The voltage across the load R_L is given by:
$$V_{dc} = I_{dc} \times R_L = \dfrac{I_m}{\pi} R_L$$

So far, it was considered that diode forward resistance is zero, but if actual resistance r_F is considered, then we get:
$$I_m = \dfrac{V_m}{(R_L + r_F)}$$

$\therefore \qquad V_{dc} = \dfrac{V_m}{\pi (R_L + r_F)} \times R_L$

Semiconductor Diodes

$$= \frac{V_m}{\pi\left(1 + \frac{r_F}{R_L}\right)}$$

or $\qquad V_{dc} = \frac{V_m}{\pi}$ for $r_F \ll R_L$.

Rectifier efficiency, $\boxed{\eta = \frac{dc \text{ power output}}{ac \text{ power input}}}$

$$= \frac{P_{dc}}{P_{ac}} = \frac{(I_m/\pi)^2 \times R_L}{(I_m/2)^2 \times (r_F + R_L)} \quad \text{as } I_{rms} = \frac{I_m}{2}$$

∴ $\qquad \eta = \dfrac{0.406 \, R_L}{r_F + R_L}$

$$= \frac{0.406}{1 + \left(\frac{r_F}{R_L}\right)}$$

or $\qquad \eta = 0.406$ for $r_F \ll R_L$.

Thus, in half-wave rectification, a maximum of 0.6% of *ac* power is converted into *dc* power.

Peak Inverse Voltage (PIV) Diode

The diode is subjected to voltage V_m in the reverse-bias situation, therefore, **peak inverse voltage (*PIV*)** in this case is V_m.

Thus, the diode must be able to withstand maximum voltage V_m in the negative half-cycle.

Example 2.5 A half-wave rectifier employs a diode having a forward resistance of 10 Ω. If the input voltage to the rectifier circuit is 12 V(rms), find the dc output voltage at a load 100 mA and PIV.

Solution:

Given: Forward resistance, $\quad r_F = 10$ W

Load Current, $\qquad\qquad\qquad I_L = 100$ mA

rms value of supply voltage, $V_{rms} = 12$ V

Maximum supply voltage, $\quad V_{SM} = \sqrt{2} \, V_{rms}$

$$= \sqrt{2} \times 12 = 16.97$$

dc output voltage for half-wave rectifier,

$$V_{dc} = \frac{V_{SM}}{\pi} - I_{dc}\, r_F$$

or

$$V_{dc} = \frac{17}{\pi} - 0.1 \times 10 = 4.4 \text{ V}.$$

$$PIV = V_{SM} = 17 \text{ V}.$$

Example 2.6 A half-wave rectifier uses a diode with an equivalent forward resistance of 0.3 Ω. If the input ac voltage is 10 V(rms) and the load is a resistance of 2.0 Ω, calculate I_{dc} and I_{rms} in the load.

Solution:

Given: Supply voltage, V_{rms} = 10 V, forward resistance,

r_F = 0.3 Ω and load resistance R_L = 20 Ω

The peak value of supply voltage,

$$V_m = 10\sqrt{2} \text{ V}$$

The peak value of current, $I_m = \dfrac{V_m}{R_L + r_F} = \dfrac{10\sqrt{2}}{2 + 0.3} = 6.15 \text{ A}$

dc output current, $\quad I_{dc} = \dfrac{I_m}{\pi} = \dfrac{6.15}{\pi} = 1.958 \text{ A}.$

rms value of output current,

$$I_{rms} = \frac{I_m}{2} = \frac{6.15}{2} = 3.075 \text{ A}.$$

2.7.2 Full-Wave Rectifier

Half-wave rectifier utilises only one half-cycle of the input wave. **Full-wave rectifier** utilises both the half cycles. A unidirectional local current is achieved by inverting alternate half cycles. Full-wave rectifier can be divided in two categories. One is known as **centre tap rectifier** which uses two diodes. The other is known as bridge rectifier which uses four diodes.

2.7.2.1 *Centre-tap rectifier*

In this case, the secondary winding is centre tapped and load along with two diodes is connected as shown in Figure 2.12(*a*). The secondary winding divided in two equal parts and a tapping is done and used in circuit as shown. The wave-form of *dc* voltage across. The load will be as shown in Figure 2.12(*b*). Diode D_1 conducts during positive half-cycle whereas diode D_2 conducts during negative half cycle. Thus, load current through load is always in one direction only. Hence, it is full-wave rectified *dc* output.

Semiconductor Diodes

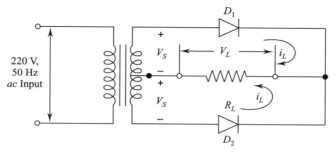

Fig. 2.12 (a) Centre-tapped full-wave rectifier.

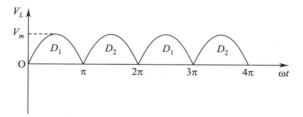

Fig. 2.12 (b) Centre-tapped full-wave rectifier wave-form.

Peak Inverse Voltage (PIV) of Diode

The voltage V_m is the maximum voltage across half of the secondary winding. When diode D_1 is conducting, resistance of diode is almost zero. Hence, full peak voltage V_m appears across the load resistor R_L. Thus, reverse voltage which appears the diode D_2 summation of voltage across D_2 and that load R_L. Hence, V_m voltage appears across diode D_2 i.e., 2 V_m voltage appears across the non-conducting diodes. D_1 or D_2.

∴ Peak inverse voltage of diode

$$PIV = 2\ V_m$$

Centre tapping is difficult, *dc* power output is small as secondary winding is divided and diodes should have high *PIV*.

2.7.2.2 Bridge rectifier

It uses four diodes instead of two as shown in Fig. 2.13. But, it does not need a centre-tapped transformer.

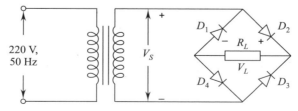

Fig. 2.13 Bridge rectifier.

A simplified circuit diagram of the **bridge rectifier** is shown in Fig. 2.14(a). Diodes D_2 and D_4 conduct during positive half-cycle of the supply whereas diodes D_1 and D_3 are non-conducting as shown in Fig. 2.14(b). Hence, current flows through the load resistor R_L, diodes D_2 and D_4. Diodes D_1 and D_3 conduct during negative half-cycle of the supply whereas diodes D_2 and D_4 are non-conducting as shown in Fig. 2.14 (c). Thus, current flows in the same direction through the load resistor R_L, diodes D_1 and D_3. Hence, an alternating bidirectional voltage wave-form is converted into unidirectional voltage wave-form across the load resistor.

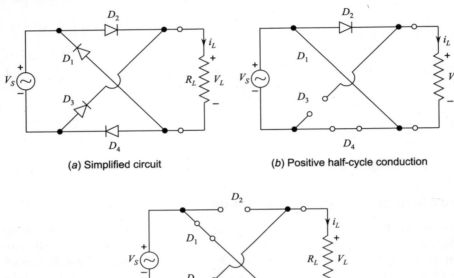

(a) Simplified circuit

(b) Positive half-cycle conduction

(c) Negative half-cycle conduction

Fig. 2.14 Simplified circuit and conduction of bridge rectifier

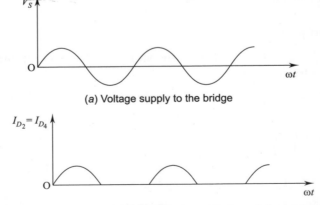

(a) Voltage supply to the bridge

(b) Current wave-form during positive half-cycle supply through the load resistor R_L

(c) Current wave-form during negative half-cycle of the supply through the load resistor R_L

(d) Net current wave-form during full-cycle of the supply through the load resistor R_L

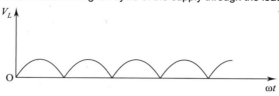

(e) Voltage wave-form during full-cycle of supply across load resistor R_L

Fig. 2.15 Bridge rectifier wave-forms.

The wave-form of the supply voltage is shown in Fig. 2.15(a). The current wave-form through the load resistor R_L during positive half cycle of the supply is shown in Fig. 2.15(b). Similarly, the current wave-form through the load resistor R_L during negative half-cycle of the supply is shown in Fig. 2.15(c). The net current wave during full cycle of the supply through the load resistor R_L is shown in Fig. 2.15(d). Thus, voltage wave-form during full cycle of supply across the load is as shown in Fig. 2.15(e), i.e., fully rectified wave-form of the supply.

The peak inverse voltage (PIV) across each non-conducting diodes in a bridge rectitifier is just the peak value of the voltage supply, i.e., V_m. Thus, the diodes used for bridge rectifier are cheaper as compared to the ones used for centre-tapped rectifiers.

It is important to note that the need for centre-tapping of supply transformer secondary is eliminated in bridge rectifier. The output is twice that of the centre tapped circuit for the same secondary voltage. For the same dc output voltage, PIV of bridge rectifier circuit is half that of centre tapped circuit. It requires four diodes, each half-cycle of ac input two diodes that conduct are in series, therefore, voltage drop in the internal resistance of the rectifying unit will be twice as great as in the centre-tapped circuit. This is undesirable when the secondary voltage is small.

2.8 RIPPLES EFFICIENCY AND REGULATION

A measure of purity of the dc output of a rectifier is ripple factor which is defined as follows:

Ripple factor, $$r = \frac{rms \text{ value of the components of wave}}{\text{average or } dc \text{ value}}$$

Rectification efficiency is defined as:

$$\eta = \frac{dc \text{ power delivered to load}}{ac \text{ input power from transformer secondary}}$$

or

$$\eta = \frac{P_{dc}}{P_{ac}}$$

It may be noted that P_{ac} is the power which would be indicated by a wattmeter connected in the rectifying circuit with its voltage terminates placed across the secondary winding and P_{dc} is the *dc* output power.

The degree of constancy is measured by load voltage regulation defined as:

$$\text{Load Regulation} = \frac{\text{No load average voltage} - \text{Full load average voltage}}{\text{Full load average voltage}}$$

2.9 EFFICIENCY OF FULL-WAVE RECTIFIER

Take $V = V_m \sin \omega t$ as the *ac* voltage given for rectification. R_L and r_F are load resistance and diode resistance respectively,

then, $P_{dc} = (I_{dc})^2 \times R_L$

for $\quad I_{dc} = \dfrac{2I_m}{\pi} \quad$ and $\quad I_m = \dfrac{V_m}{r_F + R_L}$

i.e., $\quad P_{dc} = \left(\dfrac{2I_m}{\pi}\right)^2 \times R_L$

and $\quad P_{ac} = (I_{rms})^2 \times (r_F + R_L)$

or $\quad P_{ac} = \left(\dfrac{I_m}{\sqrt{2}}\right)^2 \times (r_F + R_L) \text{ as } I_{rms} = \dfrac{I_m}{\sqrt{2}}$

Thus, $\quad \eta = \dfrac{P_{dc}}{P_{ac}} = \dfrac{(2I_m/\pi)^2 \times R_L}{(I_m/\sqrt{2})^2 \times (r_F + r_L)} = \dfrac{0.812 \, R_L}{r_F + R_L}$

or $\quad \eta = \dfrac{0.812}{1 + \left(\dfrac{r_F}{R_L}\right)}$

i.e., the efficiency will be maximum if $r_F \ll R_L$.

Hence, maximum efficiency of a full-wave rectifier is 81.2% which is double of half-wave rectifier.

2.10 FILTERS FOR RECTIFIERS

Rectifier output should be similar to a battery output. The rectifier output is pulsating *dc* which can be smoothened out using filter circuits. Figure 2.16 shows schematic of a rectifier with a shunt capacitor filter. Input and output wave-forms of the filter are also shown.

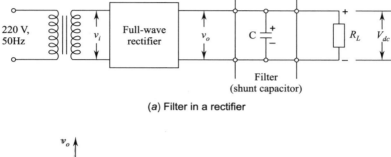

(a) Filter in a rectifier

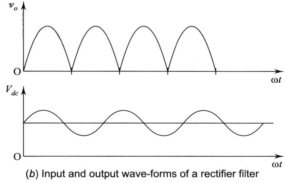

(b) Input and output wave-forms of a rectifier filter

Fig. 2.16 Full-wave rectifier with a shunt capacitance filter.

There are several types of filters which are in use, but shunt capacitor serving as a filter is most common. As seen in Fig. 2.16(a), it is basically just a large value capacitor which is connected across the full-wave rectifier and the load R_L. The pulsating input voltage is applied across the capacitor and filter output is smoothened. The capacitor changes the conditions under which the diodes conduct as shown in Fig. 2.17. When the rectifier output is increasing, the capacitor charges to peak value voltage V_m. Soon after, the rectifier voltage output tries to fall. As soon as the source voltage becomes slightly less than V_m, the capacitor will try to send current back through the diode. This reverse-biases the diode *i.e.*, it becomes open circuited. Thus, power source gets separated from the load. The capacitor starts to discharge through the load which prevents the load voltage from falling to zero. This continues to discharge until the source voltage becomes more than the capacitor voltage. This cycle keeps on repeating. The rectifier supplies the charging current through the capacitor branch as well as the load R_L. Thus, current is maintained through the load all the time at almost a constant value.

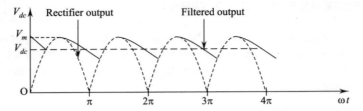

Fig. 2.17 Wave-form output of shunt capacitor filter.

Example 2.7 Sketch the output voltage v_o for the circuit given in the following figure. Assume diodes D_1 and D_2 to be ideal diodes.

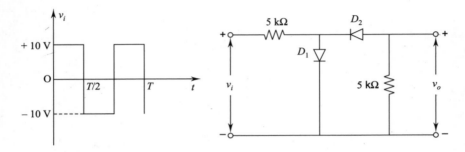

Solution: During positive half-cycles of input voltage, diode D_1 is forward biased and D_2 is reverse biased. In this case, current flows through only one diode D_1. Thus, output v_o is zero. During negative half-cycles of input voltage, diode D_1 is reverse biased and diode D_2 is forward biased, *i.e.*, current flows through only one diode D_2. The voltage v_o is given by:

Output voltage, $$v_o = \frac{v_i \times 5\ \text{k}\Omega}{(5\ \text{k}\Omega + 5\ \text{k}\Omega)}$$

or $$v_o = \frac{v_i}{2} = 5\ \text{V}$$

Thus, the output wave-form sketch is as follows:

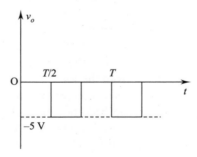

Example 2.8 Sketch the output voltage wave-form for the circuit given below. Assume the diode is ideal.

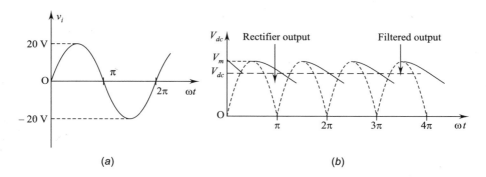

(a)   (b)

Solution:

(a) **For positive half-cycle:** The input wave-form circuit behaviour and output wave-form are as follows:

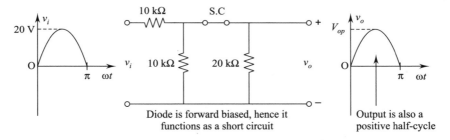

Diode is forward biased, hence it functions as a short circuit

Output is also a positive half-cycle

The peak value of output

$$V_{op} = 20 \times \frac{(10 \| 20)}{10 + (10 \| 20)} = 20 \times \frac{6.67}{10 + 6.67}$$

or $\quad V_{op} = 8.0$ volts.

(b) **For negative half-cycle:** The wave-forms and circuit are as follows:

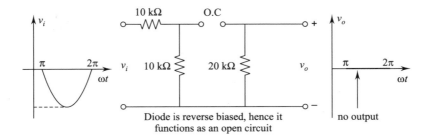

Diode is reverse biased, hence it functions as an open circuit

no output

Example 2.9 In the given circuit, calculate and sketch the wave-form of current; over one period of the input voltage. Assume the diodes to be ideal.

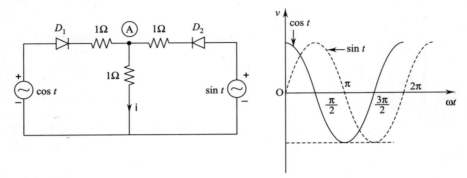

Solution:

Both D_1 and D_2 diodes conduct for $0 \leq \omega t \leq \dfrac{\pi}{2}$, where $\omega = 1$ rad./sec.

If the voltage at node is V_A, then by applying KCL, we get

$$\dfrac{V_A - \cos t}{1} + \dfrac{V_A - \sin t}{1} + \dfrac{V_A}{1} = 0$$

or $\qquad 3 V_A = \cos t + \sin t$

or $\qquad V_A = \dfrac{\cos t + \sin t}{3}$

or $\qquad i = \dfrac{V_A}{1} = \dfrac{\cos t + \sin t}{3} \qquad \ldots(i)$

During $\dfrac{\pi}{2} \leq \omega t \leq \pi$, only diode D_2 conducts as $\sin t$ is in positive half-cycle.

Thus, $\qquad i = \dfrac{\sin t}{2} \qquad \ldots(ii)$

During $\pi \leq \omega t \leq \dfrac{3\pi}{2}$, none of the diodes conduct.

$\therefore \qquad i = 0$

During $\dfrac{3\pi}{2} \leq \omega t \leq 2\pi$, D_1 conducts and D_2 does not conduct.

$\therefore \qquad i = \dfrac{\cos t}{2}.$

The output wave-form is as given below:

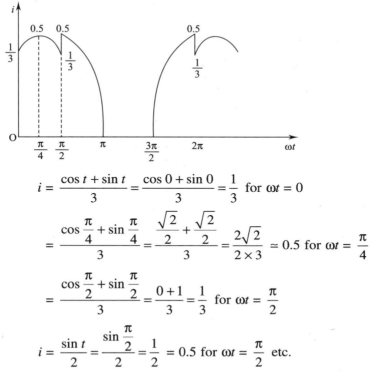

$$i = \frac{\cos t + \sin t}{3} = \frac{\cos 0 + \sin 0}{3} = \frac{1}{3} \text{ for } \omega t = 0$$

$$= \frac{\cos \frac{\pi}{4} + \sin \frac{\pi}{4}}{3} = \frac{\frac{\sqrt{2}}{2} + \frac{\sqrt{2}}{2}}{3} = \frac{2\sqrt{2}}{2 \times 3} \approx 0.5 \text{ for } \omega t = \frac{\pi}{4}$$

$$= \frac{\cos \frac{\pi}{2} + \sin \frac{\pi}{2}}{3} = \frac{0+1}{3} = \frac{1}{3} \text{ for } \omega t = \frac{\pi}{2}$$

$$i = \frac{\sin t}{2} = \frac{\sin \frac{\pi}{2}}{2} = \frac{1}{2} = 0.5 \text{ for } \omega t = \frac{\pi}{2} \text{ etc.}$$

Example 2.10 What is the ripple factor having rms value of 2 V on average of 50 V ?

Solution: rms value of ac, $V_{rms} = 2$
Average value of voltage output, $V_{ac} = 50$ V

∴ $\quad$ Ripple factor $= \dfrac{V_{rms}}{V_{dc}} = \dfrac{2}{50} = 0.04$

Example 2.11 In a power supply, the dc output voltage drops from 44 V with no-load to 42 V at full load. Calculate the percentage of voltage regulation.

Solution:

Given: No-load voltage, $\quad V_{NL} = 44$ V
Full-load voltage, $\quad V_{FL} = 42$ V

∴ $\quad$ % Voltage regulation $= \dfrac{V_{NL} - V_{FL}}{V_{FL}} \times 100 = \dfrac{44-42}{42} \times 100$

$$= \frac{2}{42} \times 100$$

$$= 4.76\%.$$

2.11 CLIPPING CIRCUITS

Diode **clipping circuits** or clippers separate an input signal at a particular *dc* level and pass the output without distortion, desired upper or lower portion of the original wave-form. Clippers are used to eliminate amplitude noise or to fabricate new wave-forms form an existing signal. There are two types of clippers—series and parallel.

A simple series clipper is a halfwave rectifier as shown in Fig. 2.18(*a*) wherein input and output wave-forms are also shown. It can be seen that the series configuration has the diode in series with load. The orientation of diode decides whether positive or negative region of the applied voltage is "clipped off". The addition of a *dc* supply to the network has pronounced effect on the clipper output, *i.e.*, can aid or work against the source voltage. The *dc* supply gives biasing effect. A biased series clippers with input and output wave-forms are shown in Fig. 2.18(*b*). A negative simple series clipper with input/output wave-forms in Fig. 2.19(*a*). A negative biased series clippers with input/output wave-forms are shown in Fig. 2.19(*b*).

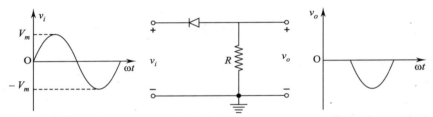

Fig. 2.18(*a*) Positive simple series clipper and input/output wave-forms.

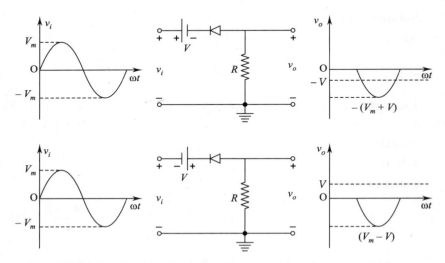

Fig. 2.18(*b*) Positive biased series clipper and input/output wave-forms.

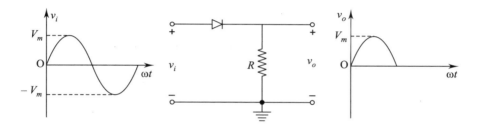

Fig. 2.19(a) A negative biased series clipper with input/output wave-forms.

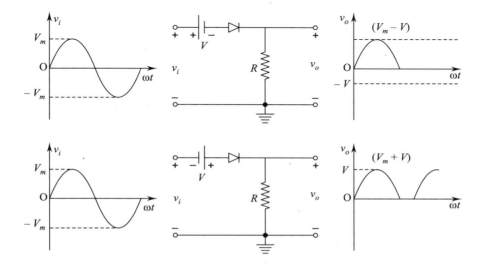

Fig. 2.19(b) Negative biased series clippers with input/output wave-forms.

A parallel clipper has the diode in a branch parallel to the load. A positive simple parallel clipper with input/output wave-forms is shown in Fig. 2.20(a). Positive biased parallel clipper with input/output wave-forms are shown in Fig. 2.20(b).

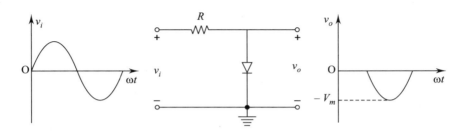

Fig. 2.20(a) A positive simple parallel clipper with input/output wave-form.

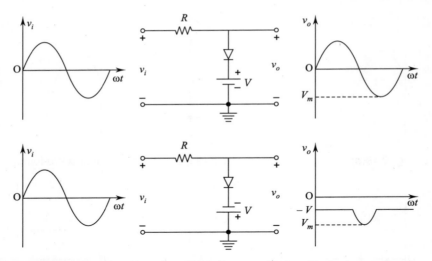

Fig. 2.20(b) Positive biased parallel clippers with input/output wave-forms.

A negative simple parallel clipper with input/output wave-forms is shown in Fig. 2.21(a). Negative biased parallel clippers with input/output wave-forms are shown in Fig. 2.21(b).

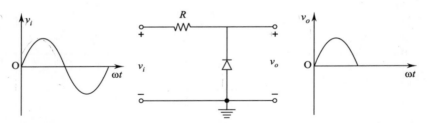

Fig. 2.21(a) A negative simple parallel clippers with input/output wave-forms.

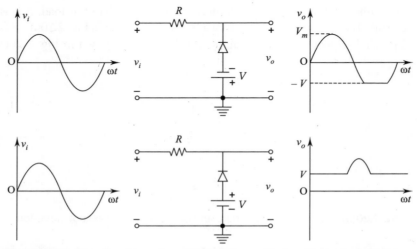

Fig. 2.21(b) Negative biased parallel clippers with input/output wave-forms.

2.12 CLAMPING CIRCUITS

A clamping circuit or **clamper** is a network made of a diode, a resistor, and a capacitor which shifts a wave-form to a different *dc* level without changing the appearance of the applied signal. Clamping circuits have a capacitor connected directly from input to output with a resistive element in parallel with output signal. Although, diode is also in parallel with the output signal but may or may not have a series *dc* supply as an added element. A simple negative clamping circuit with input/output wave-forms is shown in Fig. 2.22(*a*). Negative biased clamping circuits with input/output wave-forms are shown in Fig. 2.22(*b*).

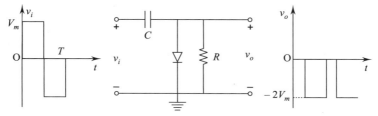

Fig. 2.22(a) A simple negative clamping circuit with input/output wave-forms.

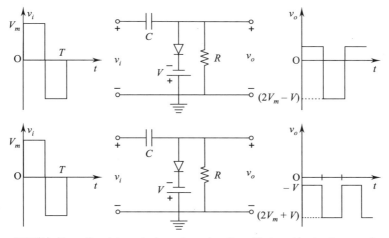

Fig. 2.22(b) Negative biased clamping circuits with input/output wave-forms.

A simple positive clamping circuit with input/output wave-forms is shown in Fig. 2.23(*a*). Positive biased clamping circuits with input/output wave-forms are shown in Fig. 2.23(*b*).

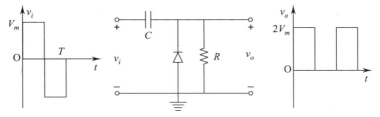

Fig. 2.23(a) A simple positive clamping circuit with input/output wave-forms.

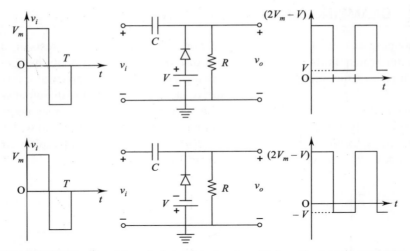

Fig. 2.23(b) Positive biased clamping circuits with input/output wave-forms.

2.13 VOLTAGE MULTIPLIERS

A **voltage multiplier** helps in giving *dc* output having multiple of peak of *ac* input. A voltage multiplier circuit is a combination of two or more peak rectifiers. Voltage multipliers can raise voltage level to hundreds or thousands of volts. Voltage multipliers are classifed as:

(*i*) Voltage doubler

    (*a*) Half-wave voltage doubler

    (*b*) Full-wave voltage doubler

(*ii*) Voltage trippler

(*iii*) Voltage quadrupler.

2.13.1 Voltage Doubler

The circuit gives *dc* output voltage which is double of the peak of *ac* input voltage. It can be half-wave voltage doubler or full-wave **voltage doubler**.

2.13.2 Half-Wave Voltage Doubler

A half-wave voltage doubler circuit is shown in Fig. 2.24. The elements D_1, C_1 and D_2, C_2 are used in the rectifier.

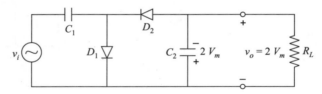

Fig. 2.24 A half-wave voltage doubler.

When during positive half-cycle D_1 conducts and D_2 is not conducting, capacitor C_1 changes upto dc peak value (V_m). But, during negative half-cycle D_2 conducts and D_1 is not conducting, hence, capacitor C_2 charges. During negative-half-cycle, the voltage across C_1 is in series with the input voltage, hence, the total voltage across capacitor C_2 is $2V_m$. Thus, capacitor C_2 charges to the voltage $2V_m$. In the next positive half-cycle, D_2 is not conducting, hence, capacitor C_2 will discharge through the load. Both the diodes D_1 and D_2 should have a peak inverse voltage (PIV) of $2V_m$ each.

2.13.3 Full-Wave Voltage Doubler

A full-wave voltage doubler circuit is shown in Fig. 2.25. During positive half-cycle diode D_1 conducts and charges capacitor C_1 to a peak voltage V_m. Diode D_2 does not conduct during this period. Diode D_2 conducts during negative half-cycle and capacitor C_2 charges to peak voltage V_m. Diode D_1 does not conduct during this period. Thus, peak voltage $2V_m$ is supplied to load R_L. The peak inverse voltage (PIV) of each diode in this case is equal to $2V_m$. It may be noted that centre-tapped transformer is not needed in this circuit.

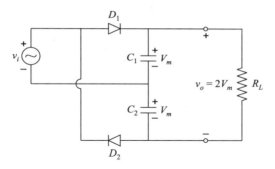

Fig. 2.25 A full-wave voltage doubler circuit.

2.13.4 Voltage Trippler and Quadrupler

A **voltage trippler** and **quadrupler circuit** is shown in Fig. 2.26. It can be seen that conduction of D_1 charges C_1 during positive half-cycle V_m peak value. Conduction of D_2 charges C_2 to peak value $2V_m$ produced by sum of source and capacitor C_1 voltage. During the second positive half-cycle D_3 will conduct and peak voltages C_1 and source V_1 will charge C_3 to $2V_m$ peak voltage. During second negative half cycle diodes D_2 and D_4 will conduct leading to C_3 charging C_4 to the same peak value $2V_m$.

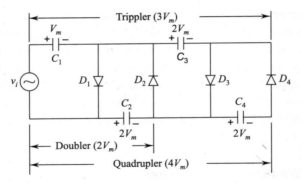

Fig. 2.26 A Voltage trippler and quadrupler circuit.

Thus, it can be seen that voltage across C_1, C_1 and C_3, C_2 and C_4 are $2V_m$, $3V_m$ and $4V_m$ respectively, *i.e.*, voltage multiplied is 2, 3 and 4. Each diode *PIV* will be $2V_m$.

2.14 ZENER DIODES

Zener diodes operate in the breakdown region without damage and these are available from about 2 V to 200 V.

There is a point where the application of too negative a voltage will result in a sharp change in the characteristics of a diode as shown in Fig. 2.27. The current increases rapidly in a direction opposite to that of positive voltage region. The reverse-bias potential which gives this dynamic change in characteristics is known as the zener potential (V_z).

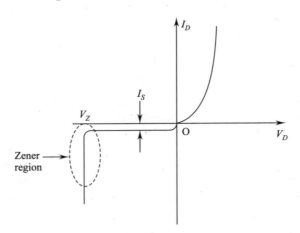

Fig. 2.27 Zener breakdown characteristics.

The breakdown or zener voltage depends upon the amount of doping. Heavy doping gives thin depletion layer and breakdown of the junction will occur at a lower reverse voltage. Light doping gives higher breakdown voltage.

Semiconductor Diodes

If the voltage of the zener-bias region increases, the reverse saturation current I_s will also increase. This aids the ionization process to the point where a high avalanche current is established which establishes avalanche breakdown region. Avalanche effect occurs due to accumulative action.

The external applied voltage accelerates the minority carriers in the depletion region. They achieve sufficient kinetic energy to ionize atoms by collision. This creates new electrons which are again accelerated to high-enough velocities to ionise more atoms. This way, an avalanche of free electrons is obtained.

Thus, reverse current increases sharply. The avalanche region depends on the doping as already discussed.

2.14.1 Zener Diode Functioning

It is a silicon junction diode which is operated under reverse-bias and arranged to breakdown when a specific reverse-bias voltage is applied. Zener diode is a crystal diode which is properly doped to have a sharp breakdown voltage. Figure 2.28 shows the **symbol of a zener diode**. It is like an ordinary diode except that the bar is turned into Z-shape.

Fig. 2.28 Zener diode symbol.

2.14.2 Zener Resistance

The current through a zener diode produces a small voltage drop in addition to the breakdown voltage. In breakdown region, variation of current through zener does not give appreciable change in the voltage drop. Hence, it is generally ignored.

Zener diode equivalent circuit is shown in Fig. 2.29. It is equivalent to a battery of voltage V_Z in series with a resistance r_Z. Resistance r_Z is called dynamic resistance or **zener resistance** of a zener diode. Zener resistance is zero for an ideal diode. The value of dynamic resistance is:

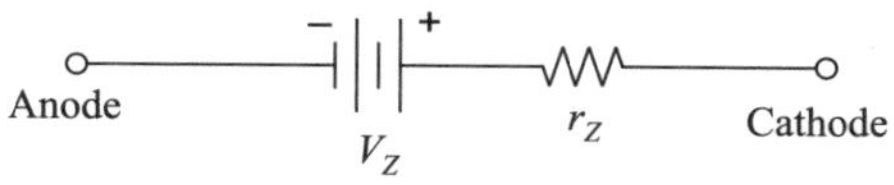

Fig. 2.29 Equivalent circuit of a zener diode.

$$r_Z = \frac{\Delta V_Z}{\Delta I_Z}.$$

The zener resistance value lies in the range of few ohms to several hundred ohms.

2.14.3 Zener 'ON' and 'OFF" States

In case reverse bias voltage across a zener diode is equal to or more than breakdown voltage V_z, the current increases sharply. The curve will be almost vertical in this region. This implies that the voltage across zener diode is constant at V_z even if the current through it changes. Thus, breakdown region of an **ideal zener diode** will be represented by a battery of voltage V_z, see Fig. 2.30. Zener diode is said to be in "ON" condition in such situation.

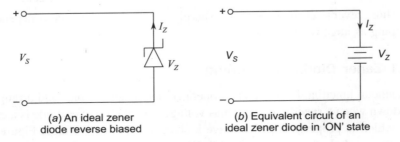

(a) An ideal zener diode reverse biased

(b) Equivalent circuit of an ideal zener diode in 'ON' state

Fig. 2.30 Zener is in 'ON' state.

In case, the reverse bias voltage across the zener diode is less than V_z but greater than zero volt, the zener diode will be in 'OFF' state. This case is represented by an open-circuit as shown in Fig. 2.31.

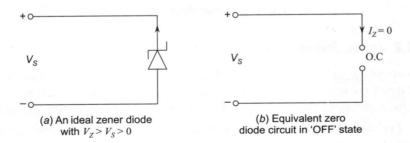

(a) An ideal zener diode with $V_Z > V_S > 0$

(b) Equivalent zero diode circuit in 'OFF' state

Fig. 2.31 Zener is in 'OFF' state.

Example 2.13 A zener diode has a breakdown voltage of 10 V in the given circuit. What are the minimum and maximum zener currents ?

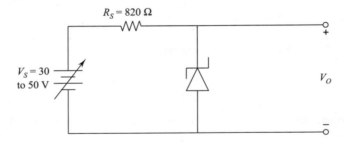

Solution: Minimum zener current, $I_s^{MIN} = \dfrac{(30-10) \text{ V}}{820 \text{ }\Omega}$ as voltage across the zener shall be 10 V to breakdown voltage.

$$\therefore \quad I_s^{MIN} = \dfrac{20}{820} = 24.4 \text{ mA}.$$

Maximum zener current, $I_s^{MAX} = \dfrac{(50-10) \text{ V}}{820 \text{ }\Omega}$

or $\quad I_s^{MAX} = \dfrac{40}{820 \text{ }\Omega} = 48.8 \text{ mA}.$

2.14.4 Zener Regulator

Zener diode maintains a constant output voltage even though the current through it as voltage **regulator**, *see* Fig. 2.32.

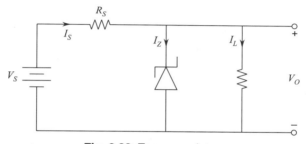

Fig. 2.32 Zener regulator.

The current through resister R_s is

$$I_s = \dfrac{V_S - V_Z}{R_S}$$

The practical zener diode will have some resistance r_z, therefore, V_L is :

$$V_L = V_z + I_z r_z$$

Normally, r_z can be neglected, hence V_L is:

$$V_L = V_z$$

The load current, $\quad I_L = \dfrac{V_L}{R_L}$

Using *KCL*, we get:

$$I_s = I_z + I_L$$
or $\quad I_z = I_s - I_L.$

This means that the zener current no longer equals the series current as it does in an unloaded **zener regulator**. Due to load resistor, the zener current equals the series current minus the load current.

2.15 TEMPERATURE COEFFICIENT

Rise in the ambient temperature leads to slight changes in the zener voltage. The effect of temperature is represented by **temperature coefficient**, which is the percentage change per degree change. Thus, calculation of zener voltage change at the highest ambient temperature is essential. The temperature coefficient is negative for zener diodes having the breakdown voltages less than 5 V. The temperature coefficient is positive for zener diodes having the breakdown voltages more than 6 V. The temperature coefficient changes from negative to positive between 5 V and 6 V. This implies that there is an operating point at which the temperature coefficient is zero. In case, the zener voltage is to be kept constant over a large temperature range in some applications, temperature coefficient is very important.

Example 2.14 A zener diode has zener voltage of 12 V and temperature coefficient $\alpha = 0.06\%/°C$. Calculate the change in zener voltage when ambient temperature of 25°C changes to 110°C.

Solution: Change in zener voltage,

$$\Delta V = V \times \frac{\alpha}{100} \times \Delta T$$

or

$$\Delta V = 12 \times \frac{0.06}{100} \times (110 - 25)$$

or

$$\Delta V = 0.61 \text{ V}.$$

Example 2.15 A zener diode has zener voltage of 3.3 V and temperature coefficient of $\alpha = -0.062\%/°C$. Calculate the zener voltage when ambient temperature of 25°C changes to 110°C.

Solution: Change in zener voltage,

$$\Delta V = V \times \frac{\alpha}{100} \times \Delta T$$

or

$$\Delta V = 3.3 \times \left(\frac{-0.062}{100}\right) \times (110 - 25)$$

or

$$\Delta V = -0.17 \text{ V}$$

∴ Zener voltage, $V = V + \Delta V$

or $V = 3.3 - 0.17 = 3.13$ V.

2.15.1 Zener Diode Ratings

The specification data of diodes are provided by the manufacturer in two forms. They give a brief description limited to one page. They may also give the characteristics using graphs, art work, tables, etc. The specifications or **ratings of diode** must include the parameters given below. Ratings of BAY 73 diode are written in the broads as an example of ambient temperature of 25°C.

(i) Forward voltage V_F at a specified current and temperature (0.60 – 0.68 V for $I_F = 1.0$ mA at 25°C).

(ii) Maximum forward current I_F at a specified temperature (500 mA at 25°C).

(iii) Reverse saturation current I_R at a specified voltage and temperature (0.5 nA, $V_R = 100$ V, $T_A = 25°C$).

(iv) Reverse-voltage rating or Peak inverse voltage (125 V at $I_R = 100$ μA).

(v) Maximum power dissipation level at a particular temperature (500 mW at 25°C).

(vi) Capacitance levels (8 pF at $V_R = 0, f = 1.0$ MHz).

(vii) Reverse recovery time t_{rr} (3 μs at $I_F = 10$ mA, $V_R = 35$ V, $R_L = 1.0$ to 100 kΩ).

(viii) Minimum reverse-bias voltage (125°C).

(ix) Operating temperature range (25°C to 125°C).

(x) Temperature coefficients.

(xi) Dimentional specifications (diagram with dimensions).

2.16 ZENER DIODE APPLICATION AS SHUNT REGULATOR

It can be used as a voltage regulator to obtain a constant voltage from a source voltage which may have large range of variation. The circuit diagram is shown in Fig. 2.33. The zener diode is reverse connected as a shunt across the load R_L. R_S

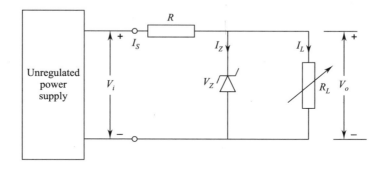

Fig. 2.33 Zener diode shunt voltage regulator.

series resistance absorbs the output voltage fluctuation such that desired constant output voltage is maintained across the load R_L. The zener will maintain a constant voltage equal to its breakdown voltage V_Z across load R_L as long as input voltage V_i does not fall below V_Z. The operating principle is as follows:

(i) If the voltage across R_L is less than the zener-breakdown voltage V_Z, then the zener diode does not conduct, i.e., resistors R_S and R_L become a potential divider across V_i.

(ii) When the V_i voltage goes above V_Z, the zener operates in the breakdown region. Resistor R_S limits the zener current from exceeding its rated maximum current.

(iii) Once zener diode conducts, the voltage across it remains almost constant although current I_Z may vary appreciably.

(iv) In case, the load current I_L increases, the current I_Z reduces to maintain current I_S constant and voltage across load R_L remains constant. Thus, the current voltage V_o remains constant.

(v) However, if the load current I_L reduces, the current I_Z increases to maintain current. Thus, the output voltage V_o remains constant.

(vi) If the input voltage V_i increases, the zener diode passes larger current so that extra voltage is dropped across resistor R_S. In case, the input voltage V_i reduces, the zener diode current also reduces and the voltage drop across R_S is reduced. Thus, the output voltage V_o remains constant. Fluctuations of V_i has very little on V_o as voltage drop across R_S is of self-adjusting nature.

2.17 DIODES FOR OPTOELECTRONICS

Optics and electronics combine to make **optoelectronics**. Some of the electronic components of optoelectronics are light-emitting diodes (LEDs), photodiodes, optocouplers, etc.

2.17.1 Light-Emitting Diodes (LEDs)

When a *PN*-junction diode is forward-biased, the potential barrier is lowered, the electrons and holes recombinations take place around the junction. Recombinations of electrons and holes radiate energy. In ordinary diodes, this energy is in the form of heat. However, semiconductor materials gallium arsenide phosphide (GaAsP) and gallium phosphite (GaP) can cause radiation of red, green or orange lights. A schematic diagram of *LED* and a seven segment display using **LEDs** are shown in Fig. 2.34.

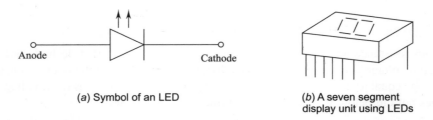

(a) Symbol of an LED

(b) A seven segment display unit using LEDs

Fig. 2.34 LED symbol and display.

2.17.2 Photodiode

A photodiode is based on principle of reverse current and it is optimised for its sensitivity to light. A window lets light pass through the package to junction. The incoming light produces free electrons and holes. Fig. 2.35 shows a circuit containing photodiode. The arrows indicate the incoming light. The reverse current is in tens of microamperes.

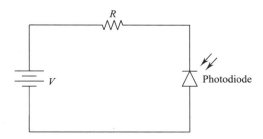

Fig. 2.35 Circuit with photodiode.

2.17.3 Optocoupler

Optocoupler or optoisolator combines a *LED* and photodiode in a single package. A circuit containing an optoccoupler is shown in Fig. 2.36. *LED* is on input side and photodiode on output side. In the input circuit, the source voltage and series resistor setup a current through *LED*. *LED* emitted light hits the photodiode which setup a reverse current in the output circuit. The reverse current produces a voltage across the output resistor. Finally, the output voltage equals the output supply voltage minus the voltage across the resistor. The output voltage varies in step with the input voltage. Thus, combination of *LED* and photodiode, *i.e.*, optocoupler transfers input signal from first circuit to second circuit. The advantage is that input and output signals are electrically isolated from each other. With an optocoupler, the only contact between two circuits is a beam of light. This gives an insulation resistance between the two circuits in thousands of mega ohms. Optocouplers are used in high-voltage applications where potentials of two circuits may differ by several thousand volts or low voltage computer signals are isolated from *ac* voltage circuits.

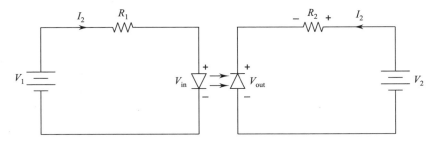

Fig. 2.36 Optocoupler circuit.

2.18 OTHER TYPES OF DIODES

Other important types of diodes are signal diodes, power diodes, Schottky diode, Varactor and Varistor. The signal diodes are normally having large reverse-resistance/forward-resistance ratio and a minimum junction capacitance. They handle small currents and/or voltages. Most types of signal diodes have a *PIV* rating in the range 30 V to 150 V. The maximum forward current range may be from 40 mA to 250 mA.

Power diodes handle large currents and/or voltages and are mostly used in rectifiers. *PIV* rating is between 50 V to 1000 V and the maximum forward current can be 30 A or even more. Power diodes are normally silicon diodes which help in reducing the voltage drop across the diode when a large forward current flows. The forward resistance is about one or two ohms. The reverse resistance is very high so that almost no current flows through the diode when reverse biased.

2.18.1 Schottky Diode

Schottky diodes are special purpose diodes which can easily rectify frequencies above 300 MHz. It does not have depletion layer which eliminates the stored charges at junction *i.e.*, switching '*ON*' and '*OFF*' is faster than ordinary diode. Schottky diodes are used in computers and infact it is a backbone of low-power *TTL* groups of devices.

2.18.2 Varactor

Varactor or varicap or epicap or tuning diode find large applications in television receivers, FM receivers, and other telecommunication receivers. It is used in reverse biased condition. The depletion layer gets wider with more reverse voltage. The *P* and *N* regions work as two plates of a capacitor. Thus, when reverse bias voltage increases, the capacitor value reduces in short, at higher frequencies, the varactor acts as a variable capacitor. Fig. 2.37 shows varactor symbol, equivalent circuit and plot of capacitance versus reverse bias voltage.

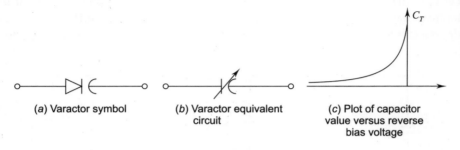

(a) Varactor symbol (b) Varactor equivalent circuit (c) Plot of capacitor value versus reverse bias voltage

Fig. 2.37 Varactor.

2.18.3 Varistor

Varistor is a diode like two back-to-back zener diodes with a high breakdown voltage in both directions. Varistor works as a transient suppressor. It protects from lightening, power-line faults which can pollute the line voltage by superimposing dips, spibes, and transients on normal 220 V supply. Dips are severe voltage drops of microseconds duration. Spibes are short overvoltages of 500 V to over 2000 V.

Example 2.16: For the circuit of the given figure, find:
(a) the output voltage
(b) the voltage drop across R_S
(c) the current through zener.

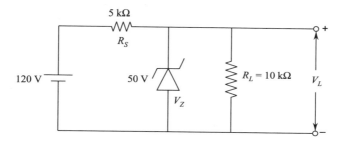

Solution:
(a) In output voltage, $V_L = V_Z = 50$ V.
(b) Voltage drop across $R_S = 120 - 50 = 70$ V.
(c) The load current, $I_L = \dfrac{50 \text{ V}}{10 \times 10^{-3} \, \Omega} = 5$ mA

Current through resistor R_S, $I_S = \dfrac{70 \text{ V}}{5 \times 10^{-3} \, \Omega} = 14$ mA

∴ Current through zener diode,
$$I_z = I_s - I_L = 14 - 5 = 9 \text{ mA}.$$

Example 2.17: For the circuit shown below, find (a) the output voltage, (b) voltage across R_S and (c) the current through zener diode.

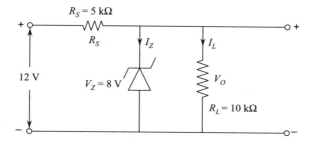

Solution:

(a) Output voltage, $V_o = V_z = 8$ V.

(b) Voltage drop across $R_S = 12 - V_o = 12 - 8 = 4$ V.

(c) Current through zener diode, $I_z = I_S - I_L$

Load current, $I_L = \dfrac{V_o}{R_L} = \dfrac{8 \text{ V}}{10 \text{ k}\Omega} = 0.8$ mA

Current through series resistor R_S, $I_S = \dfrac{12-8}{R_S}$

$= \dfrac{4 \text{ V}}{5 \text{ k}\Omega} = 0.8$ mA.

$\therefore$ Current through zener diode, $I_z = I_S - I_L$
$= (0.8 - 0.8)$ mA $= 0$.

Example 2.18 For the circuit shown below, find the maximum and minimum values of zener diode current.

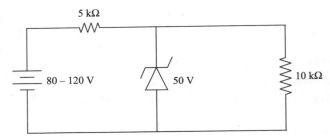

Solution: Voltage across 10 kΩ – resistor $V_o = V_z = 50$ V

Current through 10 kΩ resistor, $I_L = \dfrac{50}{10 \times 10^{-3}}$

$= 5$ mA.

Maximum Current through 5 kΩ resistor,

$$I_S^{max} = \dfrac{\text{Maximum source voltage} - V_o}{5 \text{ k}\Omega}$$

$$= \dfrac{120 - 50}{5 \times 10^{-3}} = 14 \text{ mA}$$

$\therefore$ Maximum zener current,

$$I_Z^{max} = I_S^{max} - I_L = 14 - 5 = 9 \text{ mA}.$$

Semiconductor Diodes

Minimum Current through 5 kΩ resistor,

$$I_s^{min} = \frac{\text{Minimum source voltage} - V_o}{5 \text{ k}\Omega}$$

$$= \frac{80 - 50}{5 \times 10^{-3}} = 6 \text{ mA}.$$

∴ Minimum Zener Current,

$$I_Z^{min} = 6 - 5 = 1 \text{ mA}.$$

Example 2.19 Determine V_L, I_Z and P_Z for the circuit shown below.

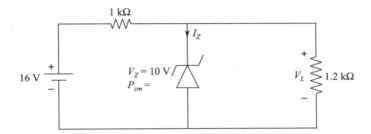

Solution: Suppose zener diode in the circuit is not conducting, then the circuit has zener diode like open. The circuit looks as follows:

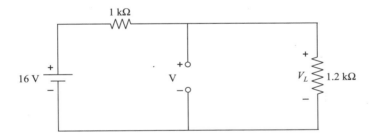

Load voltage, $$V_L = \frac{16 \text{ V} \times 1.2 \text{ k}\Omega}{1 \text{ k}\Omega + 1.2 \text{ k}\Omega}$$

or $$V_L = \frac{16 \times 1.2}{2.2} = 8.73 \text{ V}.$$

The load voltage V_L is less than zener breakdown voltage $V_Z = 10$ V, therefore, zener diode does not conduct at all.

Thus, $$I_Z = 0.$$

and $$P_Z = I_Z \times V_Z = 0 \times 8.73 = 0 \text{ }\Omega.$$

SUMMARY

1. **Diode Symbol:** The diode symbol looks like an arrow which points in the easy direction of conventional current flow. The opposite way is the easy direction for electron flow. The *P*-side is known as anode, and the *N*-side is known as cathode.

2. **Diode Characteristics:** *V-I* characteristics is a plot of forward-bias and reverse-bias currents versus external voltage applied across the diode. The characteristics is non-linear. Very high current flows in the forward-biased diode and only a small current flows in a reverse-biased diode.

3. **Knee Voltage:** The forward region has a segment known as knee voltage. This voltage is approximately equal to the barrier potential of the diode. A current-limiting resistor is always used to prevent the current from exceeding the maximum rating.

4. **Diode Current Equation:** The diode current equation is given by:

$$I_F = I_S \left(e^{-\frac{V_F}{\eta V_T}} - 1 \right)$$

where, I_F = forward diode current

I_S = reverse diode current at room temperature

V_F = external voltage applied to the diode

η = a constant, 1 for Ge and 2 for Si

and the voltage equivalent of temperature is given by:

$$V_T = \frac{kT}{q}$$

where, k = Boltzmann's constant = 1.38×10^{-23} J/K

q = electronic charge = 1.6×10^{-19} Coulomb

T = diode junction temperature in (°K)

5. **Ideal Diode:** The ideal diode is visualised as a switch which automatically closes when forward-biased and opens when reverse-biased.

$$R = \frac{V_S - V_F}{I_F}$$

6. **Forward Resistance:**

$$r_F = \frac{\eta V_T}{I_F}$$

Semiconductor Diodes

7. **Diode Approximation:**

$$V_F = V_T + I_F\, r_F$$

where, V_T = barrier potential, 0.7 V for Si, 0.3 V for Ge and 1.2 V for GaAs.

8. **Current Limiting Resistor:**

$$R = \frac{V_S - V_F}{I_F}$$

9. **Forward Current Approximation:**

$$I_F = \frac{V_S - V_B}{R + r_F}.$$

10. **Data Sheet:** It specifies the characteristics of semiconductor devices. The data sheet of a diode contains useful informations such as breakdown voltage, maximum forward current, forward voltage drop, and maximum reverse current.

11. **Input Transformer:** A step-down transformer is used in rectifiers. It may be centre-tapped in secondary winding.

12. **RMS Voltage:**

$$V_{rms} = 0.707\, V_m.$$

13. **Half-Wave Rectifier:** It has a diode in series with a load resistor. The load voltage is a half-wave rectified sine wave with a peak value approximately equal to the peak secondary voltage.

$$V_{dc} = 0.318\, V_m.$$

14. **Full-Wave Rectifier:** It has a centre-tapped transformer with two diodes and a load resistor. The load voltage is a full-wave rectified sine wave with a peak value approximately equal to half of the peak secondary voltage.

$$V_{dc} = 0.636\, V_m.$$

15. **Bridge Rectifier:** It has four diodes. The load voltage is a full-wave rectified sine wave with a peak value approximately equal to peak secondary voltage.

$$V_{dc} = 0.636\, V_m$$

$$f_{out} = 2 f_{in}$$

16. **Capacitor-Input Filter:** It is a capacitor across the load resistor which charges to the peak voltage and supplies the current to the load when the diodes are not conducting. A large capacitor gives small ripple and the load voltage is almost a pure *dc* voltage.

17. **Diode Current for Full-Wave Rectifier:**

$$I_F = 0.5\, I_L$$

18. **Peak Inverse Voltage (PIV):**

For a full-wave rectifier:

$$PIV = \text{peak secondary voltage of the transformer.}$$

19. **Ripple Factor (r):**

$$r = \sqrt{\left(\frac{I_{rms}}{I_{dc}}\right)^2 - 1}$$

20. **Clipper:** A circuit which shapes the wave-form by removing or clipping a portion of the applied input signal wave-form without distorting the remaining part.

21. **Chamber:** A circuit which shifts or clamps a signal to different *dc* level.

22. **Voltage Multiplier:** A circuit which produces *dc* output whose value is multiple of peak *ac* input voltage.

23. **Zener Diode:** A special diode optimised for operation in the breakdown region. It is used as voltage regulator.

24. **Loaded Zener Regulator:** Zener diode is connected in parallel with a load resistor. The current through the current limiting resistor equals the sum of the zener current and the load current.

25. **Optoelectronic Diodes:** Light-Emitting Diodes (*LED*) radiate light when in breakdown condition. These are used as indicators. By combining seven *LED*s in a package, a seven segment indicator / display is created.

    Photodiode is another optoelectronic diode. It is optimised for its sensitivity to light. Optocoupler is combination *LED* and photodiode in a light package which serves as electrical isolator for sensitive circuits.

26. **Schottky Diode:** It is a special diode which is useful at high frequencies where short switching times are needed. It is used generally for frequencies above 300 MHz.

27. **Varactor:** The reverse biased *P-N* junction work like a plate of capacitor. The capacitance so created is varied by controlling the reverse voltage.

28. **Varistor:** It is like two back-to-back zener diodes with a high breakdown voltage in both directions. It is used as a spibe suppressor.

EXERCISES

2.1. Explain how a barrier potential is developed at the P-N junction.

2.2. Explain the action of P-N junction diode under forward bias and reverse bias.

2.2. Draw and explain the V-I characteristics of a P-N junction diode.

2.4. Differentiate between transition capacitance and diffusion capacitance of a P-N junction diode.

2.5. What is static resistance of diode ? How will you find the dynamic resistance ?

2.6. Explain the formation of potential barrier in P-N junction. Why is silicon preferred to germanium in the manufacturing of semiconductor device ?

2.7. For a semiconductor diode, define static and dynamic resistance.

2.8. Plot the hole current, the electron current, and the total current as a function of distance on both the sides of a P-N junction. Indicate the transition region.

2.9. When a reverse bias is applied to a germanium diode, the reverse saturation current at room temperature is 0.3 µA. What is the value of current flowing in the diode when 0.15 V forward bias is applied at room temperature ?   [**Ans.** 120.73 µA]

2.10. A semiconductor diode has a forward 72 bias of 200 mV and reverse saturation current of 1 µA at room temperature. Find *ac* resistance of the diode.   [**Ans.** 11.86 Ω]

2.11. The figure given below shows the circuit of series diode configuration. What is the value of V_F, V_R and I_F ?   [**Ans.** 0.7 V, 7.3 V, 3.65 mA]

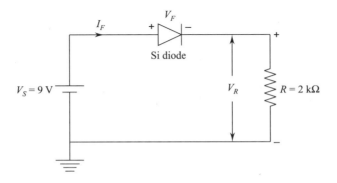

2.12. A series diode configuration circuit is given below. What is the value of V_R and I_F? **[Ans.** 11 V, 2.5 mA**]**

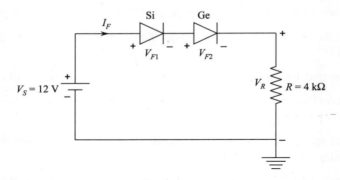

2.13. A series diode configuration circuit is given. What are the values of V_{R1}, V_{R2} and V_o? **[Ans.** 2.38 mA, 9.52 V, 4.76 V, −0.24 V**]**

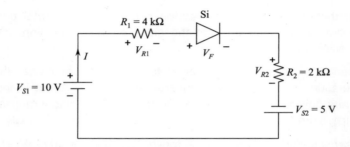

2.14. A parallel diode configuration circuit is shown in the figure given below. What are the values of V_o, I, I_{F1} and I_{F2}? **[Ans.** 0.7 V, 33.1 mA, 16.5 mA, 16.5 mA**]**

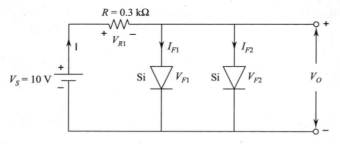

2.15. What is a resistance for a semiconductor diode with a forward bias of 0.25 V? Reverse saturation current at room temperature is of 1.2 μA.

[Ans. 1.445 Ω**]**

Semiconductor Diodes

2.16. Determine the current flowing in the circuit shown below.

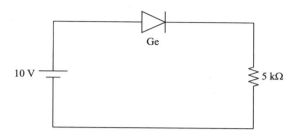

2.17. Determine the current through 2 kΩ resistor in the circuit given below.

[**Ans.** 1.205 mA]

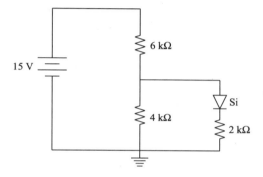

2.18. A half-wave rectifier uses a diode with an equivalent forward resistance of 0.3 Ω. If the input *ac* voltage is 10 V (rms) and the load is a resistance of 20 Ω, calculate I_{dc} and I_{rms} in the load. [**Ans.** 1.958 A, 3.075 A]

2.19. A zener diode shunt regulator circuit is shown in the figure given below. Find the zener current for the load resistances of 30 kΩ, 10 kΩ and 3 kΩ.

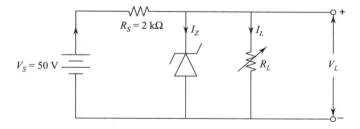

2.20. A 10 V regulated *dc* supply of 10 mA is required from a *dc* source of 12-15 V by using a pair of zener diodes. Take $I_{Z\,min} = 0.2$ mA, during the circuit, find the value of R_S and power rating of zener diodes.

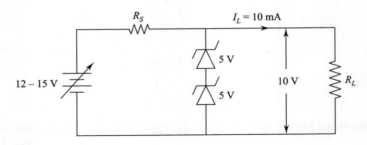

[**Ans.** $R_s = 196\ \Omega$, Power rating of zener = 255 mW]

BIPOLAR JUNCTION TRANSISTOR (BJT)

3.1 BASIC CONSTRUCTION

Bipolar Junction Transistor (BJT) was invented by Shockley in 1951 to amplify radio and TV signals. BJT replaced the vacuum tube which needed heater for its internal filament requiring, watt power. The word bipolar is an abbreviation for "two polarities". The word transistor is an acronym, and is a combination of the words "Transfer Varistor" used to their mode of operation way back in their early days of development. We know that simple diodes are made up from two pieces of semiconductor either Silicon or Germanium to form a simple P-N junction. If we join together two individual diodes end to end giving two P-N junctions combined together in series, we now have a three layer, two junction, three terminal device forming the basic Bipolar Junction Transistor (BJT) as shown in Fig. 3.1. Thus, a construction of a transistor has three doped regions giving three terminals.

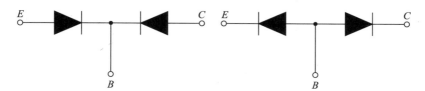

Fig. 3.1 Two individual diodes joined together.

Outer two terminals are known as emitter and collector whereas the middle terminal is known as base; there are two basic types of transistor construction, PNP and NPN, which basically describes the physical arrangement of the *P*-type and *N*-type semiconductor materials from which they are made. Transistor construction is shown in Fig. 3.2. **Transistors** are current amplifying and regulating devices. The principle of operation of the two transistor types PNP and NPN, is exactly the same only difference being in the biasing (base current) and the polarity of power supply for each type. Circuit symbols for both PNP and NPN transistors are shown in Fig. 3.3. The arrow in the circuit symbol always shows the direction of conventional current flow between the base terminal and its emitter terminal. The direction of

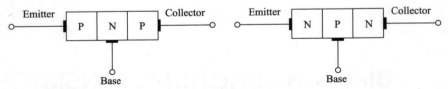

Fig. 3.2 PNP and NPN Tranistors.

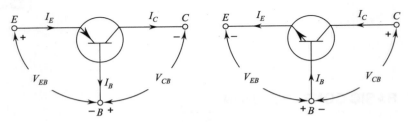

Fig. 3.3 Symbolic form.

the arrow points from the positive P-type region to the negative N-type region, exactly same as for the standard diode symbol. Current direction can also be derived by treating terminals as PNP (Positive-Negative-Positive) and NPN (Negative-Positive-Negative).

3.2 TRANSISTOR ACTION

The emitter base junction is forward biased whereas collector base junction is reverse biased for active operating as shown in Fig. 3.4. The forward-bias of the

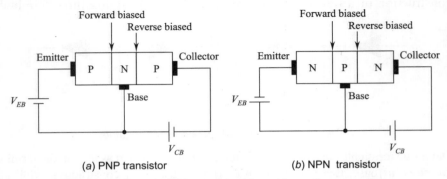

Fig. 3.4 PNP and NPN transistor biased for active operation.

emitter base causes the emitter current to flow. It can be observed that almost all emitter current flows in the collector circuit. Thus, current in the collector circuit depends on the emitter current, *i.e.*, when the emitter current is zero, the collector current is almost zero. Suppose, the emitter current is 1 mA, then the collector current is also near about 1 mA. This is the basic function of a transistor.

3.2.1 Working of PNP Transistor

Consider that a **PNP transistor** has a forward-bias on emitter-base and reverse-bias on collector-base junction as shown in Fig. 3.5. Forward bias on emitter base causes the holes in P-type emitter to flow towards the base. The holes cross into the N-type base which constitutes the emitter current I_E. They have tendency to combine with the electrons. The base is lightly doped and also very thin, hence, just few holes (less than 5%) combine with electrons. Remaining (more than 95%) cross into collector region to constitute collector current I_C. Thus, almost all the emitter current flows into the collector circuit. It should be noted that current flow within PNP transistor is due to movement of holes, although current in external wires is by electrons. It can be observed that the emitter current is the sum of collector and base currents i.e., $I_E = I_B + I_C$.

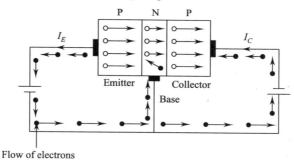

Fig. 3.5 Biased PNP transistor.

3.2.2 Working of NPN Transistor

Consider that a **NPN transistor** has a forward-bias on emitter-base and reverse-bias on collector-base junction as shown in Fig. 3.6. Due to forward-bias, electrons in N-type emitter flow towards the base. This constitutes the emitter current I_E. These electrons flow through the P-type base where they tend to combine with holes. The base is lightly doped and also is very thin, hence, just a few electrons (less than 5%) combine with the holes to constitute base current I_B. Remaining electrons (more than 95%) crossover into the collector region which constitutes collector current I_C. It is clear that: $I_E = I_B + I_C$.

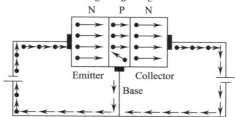

Fig. 3.6 Biased NPN transistor.

3.3 CIRCUIT CONFIGURATIONS

There are three possible configurations to connect a transistor within an electronic circuit. Each configuration responds differently as the characteristics vary with each type of circuit arrangement with regards to the input signals. The configurations are common base having voltage gain but no current gain, common emitter having both voltage gain and current gain; and common collector having current gain but no voltage gain.

3.3.1 Common Base (CB) Configuration

Common base or grounded configuration is shown in Fig. 3.7. The base is common to both the input signal and output signal. The input signal is applied between base and emitter terminals. The output signal is taken from between the base and the collector terminals. The base terminal is grounded or connected to a fixed reference voltage point. The current flowing through the emitter is quite large as base and emitter junction is forward biased. The flow through collector is very small as base and collector junction is reverse biased. Thus, the collector current output is less than the emitter current input resulting in current gain less than one, *i.e.*, it "attenuates" the input signal. Common base configuration is non-inverting voltage amplifier circuit with input signal voltage and output voltage being on phase. This configuration is not very common due to its usually high voltage characteristics. It *has a high output to input resistance*. The important aspect is that load resistance (R_L) to input resistance (R_{in}) gives gain value of "resistance gain". Thus, the voltage gain for common base can be written as:

$$A_V = \frac{I_C \times R_L}{I_E \times R_{IN}} = \alpha \times \frac{R_L}{R_{IN}} \qquad ...(3.1)$$

The common base configuration is normally used in single state amplifier circuits such as microphoto amplifier or *RF* radio amplifiers as it has very good high frequency response.

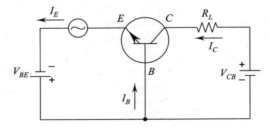

Fig. 3.7 Common base amplifier circuit.

3.3.2 Common Emitter (CE) Configuration

Common emitter or grounded emitter configuration in shown in Fig. 3.8. The input signal is applied between base and emitter while output signal is picked up between collector and emitter. This configuration is generally used for transistor based current amplifiers it produces the highest voltage current and power gain of all of the three transistor configurations. The input impedence is low as the terminals are forward biased and output impedance is high as the terminals are reverse-biased.

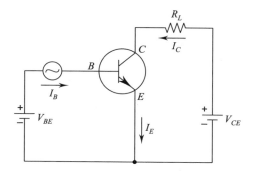

Fig. 3.8 Common emitter amplifier circuit.

From the circuit of Fig. 3.5, we know:

$$I_E = I_C + I_B$$

where I_E is current flowing out of amplifier, I_C is current flowing into the collector and I_B is the current flowing into the base and the current gains are given by:

$$\alpha = \frac{I_C}{I_E} \qquad \ldots(3.2)$$

and

$$\beta = \frac{I_C}{I_B} \qquad \ldots(3.3)$$

or

$$\alpha = \frac{\beta}{\beta+1} \qquad \ldots(3.4)$$

and

$$\beta = \frac{\alpha}{1-\alpha} \qquad \ldots(3.5)$$

In short, common emitter configuration has greater input impedance, current and power gain than that of the common base configuration but its voltage gain is much lower. It is an inverting amplifier circuit resulting in the output signal being 180° out of phase with the voltage signal.

3.3.3 Common Collector (CC) Configuration

Common collector or grounded collector configuration is shown in Fig. 3.9. The collector is common and the input signal is connected to the base and the output is taken from the emitter load. This configuration is known as a voltage follower or

emitter follower circuit. This configuration is very useful for impedance matching applications because of its high input impedance in the range of hundreds of thousands of ohms, and its output impedance is relatively low.

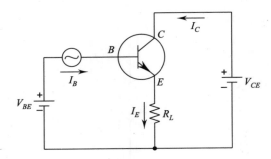

Fig. 3.9 Common Collector amplifier circuit.

The current and gain relationships are as follows:

$$I_E = I_C + I_B$$

and

$$A_i = \frac{I_E}{I_B} = \frac{I_C + I_B}{I_B}$$

or

$$A_i = \frac{I_C}{I_B} + 1$$

or

$$A_i = \beta + 1 \qquad \ldots(3.6)$$

The common collector configuration is a non-inverting amplifier circuit wherein the input signal is in phase with the output signal. In this case voltage gain is always less than unity and provides good current amplification.

3.4 INPUT/OUTPUT CHARACTERISTICS

Characteristic curves give complete behaviour of a transistor. These are relationships of transistor current and voltages. The relationship of input current and input voltage (for a value of output voltage) is known as input characteristics. The relationship of output current and output voltage for a value of input current is known as output characteristics.

3.4.1 CB Characteristics

The input characteristics for the common-base amplifier is shown in Fig. 3.10 for a silicon transistor. It relates input current (I_E) to an input voltage (V_{BE}) for various levels of output voltage.

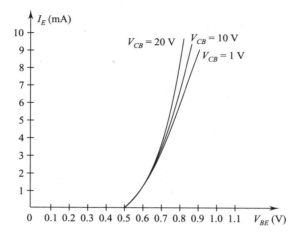

Fig. 3.10 Input characteristics of CB transistor (silicon) amplifier.

The output characteristics for the common-base amplifier is shown in Fig. 3.11. It relates output current (I_C) to an output voltage (V_{CB}) for various levels of input current (I_E).

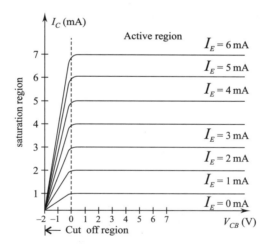

Fig. 3.11 Output characteristics of CB transistor amplifier.

The output characteristics have three basic regions known as active, cut-off and saturation as shown in the figure. **Active region** is used for linear amplifiers. In this case base-emitter is forward-biased and collector-base is reverse-biased.

In the **cut-off region**, the collector (I_C) current is zero. In this case, the base-emitter and collector-base junctions are both reverse-biased. The active, cut-off in the **saturation region**, V_{CB} is less than zero. The collector current jumps to high values as it approaches $V_{CB} = 0$. In this case, the base-emitter and collector-base junctions are both forward-biased.

Current amplification factor, $\alpha = \dfrac{\Delta I_C}{\Delta I_E}$ (α = 0.9 to 0.99)

Total Collector current, $I_C = \alpha I_E + I_{CBO}$ ...(3.7)

where I_{CBO} = Collector to base current where emitter is open very small in mA, so may be ignored.

3.4.2 CE Characteristics

The input characteristics of the common-emitter amplifier is shown in Fig. 3.12 for a silicon transistor. It relates input current (I_B) to an input voltage (V_{BE}) for various levels of output voltage. Input resistance,

$$r_i = \dfrac{\Delta V_{EB}}{\Delta I_E} \text{ at constant } V_{CB}.$$

(very small few ohms) V_{CB} and output resistance,

$$r_o = \dfrac{\Delta V_{CB}}{\Delta I_C} \text{ at constant } I_E \text{ (very high in terms of k}\Omega\text{)}.$$

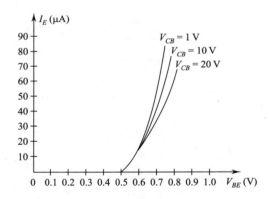

Fig. 3.12 Input characteristics of CE transistor (silicon) amplifier.

The output characteristics for the common-emitter amplifier is shown in Fig. 3.13. It relates output current (I_C) to an output voltage (V_{CE}) for various levels of input current (I_B) and saturation regions are marked.

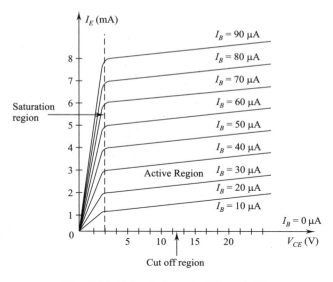

Fig. 3.13 Output characteristics of CE.

In the active region of common-emitter amplifier, the base-emitter junction is forward biased and the collector-emitter junction is reverse-biased.

Current amplification factor,

$$\beta = \frac{\Delta I_C}{\Delta I_B} \quad (20 - 500)$$

For *dc* values, $\quad \beta_{dc} = \frac{I_C}{I_B}.$

Input resistance, $\quad r_i = \frac{\Delta V_{BE}}{\Delta I_B}$ at constant V_{CE} (few hundred ohms)

Output resistance, $\quad r_o = \frac{\Delta V_{CE}}{\Delta I_C}$ at constant I_B (in the order of 50 kΩ)

Callector current $\quad I_C = \beta I_B + I_{CEO}$

Where $\quad I_{CEO}$ = Collector to Emitter current when base is open

3.4.3 CC Characteristics

Common-collector configuration is used for impedance matching as it has a high input impedance and low output impedance, *i.e.*, opposite to that of the CB and CE configurations. There is no need of CC characteristics from design point of view. The output characteristics of CC (I_E versus V_{CE}) for a range of values I_B, are same as for CE configuration with $I_E \simeq I_C$ (as $\alpha \simeq 1$) and V_{CE} = negative of V_{CE} (in CE configuration). Input characteristics of CC configuration common-emitter base characteristics are capable of giving the required information.

3.5 MATHEMATICAL RELATIONSHIPS

3.5.1 Relation between β and α

Transistor current relationships are shown in Fig. 3.14.

We know, $\qquad I_E = I_B + I_C$

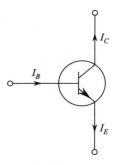

Fig. 3.14 Transistor current relationships.

or $\qquad \Delta I_E = \Delta I_B + \Delta I_C$

but $\qquad \beta = \dfrac{\Delta I_C}{\Delta I_B} = \dfrac{\Delta I_C}{\Delta I_E - \Delta I_C}$

$\qquad = \dfrac{\Delta I_C / \Delta I_E}{(\Delta I_E / \Delta I_E) - (\Delta I_C / \Delta I_E)}$

or $\qquad \boxed{\beta = \dfrac{\alpha}{1-\alpha}}$ as $\boxed{\alpha = \dfrac{\Delta I_C}{\Delta I_E}}$

i.e., if $\alpha \to 1$, $\beta \to \infty$. Thus, current gain in common emitter is very high.

3.5.2 Relation between I_{CEO} and I_{CBO}

We know, $\qquad I_C = \alpha I_E + I_{CBO}$

$\qquad\qquad\quad = \alpha(I_B + I_C) + I_{CBO}$

or $\qquad (1 - \alpha) I_C = \alpha I_B + I_{CBO}$

or $\qquad I_C = \dfrac{\alpha}{1-\alpha} I_B + \dfrac{I_{CBO}}{1-\alpha}$ ...(3.8)

and $\qquad I_C = \beta I_B + I_{CEO}$ ...(3.9)

From (3.8) and (3.9), we get:

$$I_{CEO} = \dfrac{I_{CBO}}{1-\alpha} \qquad ...(3.10)$$

3.6 BIASING OF TRANSISTORS

Biasing of a transistor is application of dc voltages to establish a fixed level of current and voltage. It is also known as operating point or quiescent (Q) point. Consider use of a transistor as an amplifier. Normal requirement is that the output should be a faithful amplification of input signal without any change in the shape. Fig. 3.15 shows conditions of proper biasing of the common-base amplifier in the active region where in $I_C \simeq I_E$, i.e., $I_B \simeq 0$ i.e., as though base is open circuited. The dc supplies are then incorporated with a polarity that will support the resulting current direction.

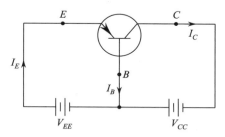

Fig. 3.15 Biasing condition of common-base PNP transistor in active region.

Fig. 3.16 shows conditions of proper biasing of a common-emitter amplifier in the active region. The dc supplies are incorporated with polarities which will support the established direction of I_E as per arrow of the transistor and other currents as per Kirchoff's current law, i.e.,

$$I_C + I_B = I_E$$

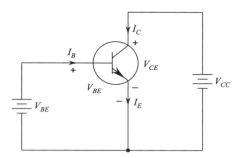

Fig. 3.16 Proper biasing condition of a common emitter NPN transistor.

The important relationship in each configuration is given by:

$$V_{BE} = 0.7 \text{ V}$$
$$I_E = (\beta + 1) I_B \simeq I_C$$
$$I_C = \beta I_B \qquad \qquad ...(3.11)$$

3.6.1 Fixed-Bias

It is the simplest transistor dc bias configuration as shown in Fig. 3.17. The capacitors in the circuit are treated as open for *dc* analysis. *i.e.*

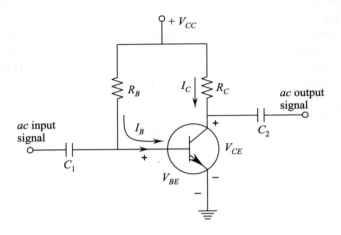

Fig. 3.17 Fixed-bias circuit of a NPN transistor.

$X_C = \dfrac{1}{2\pi f C} = \infty$ as $f = 0$ for *dc* analysis. The base-emitter loop gives:

$$V_{CC} - I_B R_B - V_{BE} = 0$$

or
$$I_B = \dfrac{V_{CC} - V_{BE}}{R_B} \qquad ...(3.12)$$

The collector-emitter loop gives:
$$I_C = \beta I_B$$

It is also important to remember that the collector current is not dependent on the load in the active region. By KVL in the collector-emitter loop, we get:

$$V_{CE} + I_C R_C - V_{CC} = 0$$

or
$$V_{CE} = V_{CC} - I_C R_C.$$

Consider the circuit of Fig. 3.17. It gives the equation for load line:

$$V_{CE} = V_{CC} - I_C R_C$$

or
$$I_C = \dfrac{V_{CC} - V_{CE}}{R_C} \qquad ...(3.13)$$

The load line is superimposed on the output characteristics as shown in Fig. 3.18.

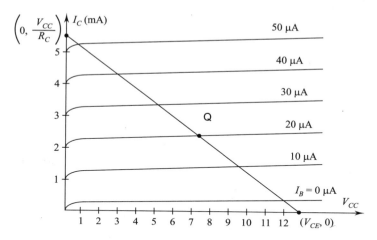

Fig. 3.18 Fixed-bias load line.

We choose the curve for I_B current (say 20 mA) and the intersection point with load line as Q. Point Q is the operating point.

3.6.2 Emitter Bias

A dc bias circuit is shown in Fig. 3.19. It contains an emitter resistance to improve the stability level over that of the **fixed-bias** configuration.

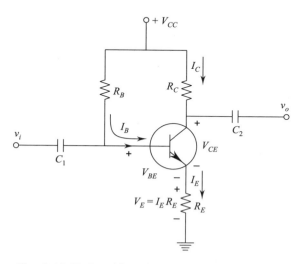

Fig. 3.19 Emitter bias circuit with emitter-resistor.

By applying KVL in base-emitter loop, we get:

$$V_{CC} - I_B R_B - V_{BE} - I_E R_E = 0$$

We know that:

$$I_E = (\beta + 1) I_B$$

By substituting value of I_E in the loop equation we get:

$$V_{CC} - I_B R_B - V_{BE} - (\beta + 1) I_B R_E = 0$$

or
$$I_B = \frac{V_{CC} - V_{BE}}{R_B + (\beta + 1) R_E} \qquad ...(3.14)$$

By applying KVL in collector-emitter loop we get:

$$I_E R_E + V_{CE} + I_C R_C - V_{CC} = 0$$

By taking $I_E \simeq I_C$, we get:

$$V_{CE} = V_{CC} - I_C (R_C + R_E) \qquad ...(3.15)$$

The voltage from emitter to ground is:

$$V_E = I_E R_E$$

and voltage from collector to ground is:

$$V_C = V_{CE} - V_E$$

or
$$V_C = V_{CC} - I_C R_C \qquad ...(3.16)$$

The voltage from base to ground is:

or
$$V_B = V_{BE} + V_E \qquad ...(3.17)$$

Saturation Level

The maximum collector current or the collector **saturation level** for an **emitter-bias** circuit can be determined by applying short circuit between collector and emitter is $V_{CE} = 0$, we get:

$$I_{C\,sat} = \frac{V_{CC}}{R_C + R_E} \qquad ...(3.18)$$

Load-line analysis of the improved bias stability will be same except that it will follow the equation:

$$I_C = \frac{V_{CC} - V_{CE}}{R_C + R_E} \qquad ...(3.19)$$

3.6.3 Voltage-Divider Bias

The bias current I_{CQ} and V_{CEQ} are dependent on parameter b which is temperature sensitive, particularly for silicon transistors. **Voltage divider** bias circuit as shown in Fig. 3.20, is independent of b parameter as change in b is very small. The analysis of the circuit can be exact or approximate.

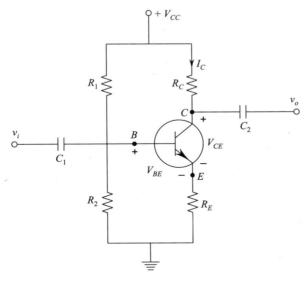

Fig. 3.20 Voltage-divider bias circuit.

3.6.3.1 *Exact analysis*

The supply between base and ground *i.e.*, input side can be represented as shown in Fig. 3.21.

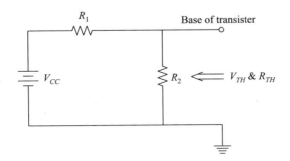

Fig. 3.21 Input side of the transistor.

Thevenin equivalent values are:

$$R_{TH} = R_1 \parallel R_2, \ (V_{CC} \text{ is taken as short circuit})$$

$$V_{TH} = \frac{R_2 V_{CC}}{R_1 + R_2} \qquad \ldots(3.20)$$

The input portion of circuit can be redrawn along with the transistor as shown in Fig. 3.22.

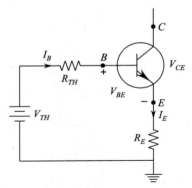

Fig. 3.22 Input portion circuit using Thevenin equivalent.

By applying KVL, we get:

$$V_{TH} - I_B R_{TH} - V_{BE} - I_E R_E = 0$$

The above equation after substituting $I_E = (\beta + 1) I_B$ gives:

$$I_B = \frac{V_{TH} - V_{BE}}{R_{TH} + (\beta + 1) R_E} \qquad ...(3.21)$$

It can be observed that R_{TH} is large, effect of β is very much reduced.

By applying KVL from V_{CC} to ground through collector and emitter circuit, we get:

$$V_{CE} = V_{CC} - I_C (R_C + R_E) \qquad ...(3.22)$$

3.6.3.2 Approximate analysis

The input circuit along with base to emitter and ground can be represented as shown in Fig. 3.23.

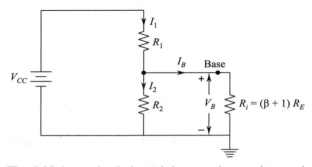

Fig. 3.23 Input circuit through base emitter and ground.

Now, if $R_i \gg R_2$, then $\quad I_B \simeq 0$, i.e., $I_1 \simeq I_2$.

Hence, $\qquad V_B = \dfrac{R_2 \cdot V_{CC}}{R_1 + R_2} \qquad ...(3.23)$

Further, $R_i = (\beta + 1) R_E \simeq \beta R_E$.

∴ The condition for R_1 to be very large than R_2 is taken as:

$$\beta R_E \geq 10 R_2$$

We can also conclude:

$$V_E = V_B - V_{BE}$$

$$I_E = \frac{V_E}{R_E}$$

$$I_{CQ} \simeq I_E$$

$$V_{CE} = V_{CC} - I_C R_C - I_E R_E$$

Taking $I_E \simeq I_C$, we get:

$$V_{CEQ} = V_{CC} - I_C (R_C + R_E) \qquad \ldots(3.24)$$

Transistor saturation equation is given by:

$$I_{C\,sat} = I_{C\,max} = \frac{V_{CC}}{R_C + R_E}$$

The load line equation is given by:

$$I_C = \frac{V_{CC}}{R_C + R_E}\bigg|_{V_{CE} = 0}$$

and

$$V_{CE} = V_{CC}\big|_{I_C = 0\,mA}$$

3.6.4 DC Bias with Voltage Feedback or Collector Bias

A feedback path from collector to base improves level of stability. Consider circuit of Fig. 3.24 which is having a **voltage feedback** from collector.

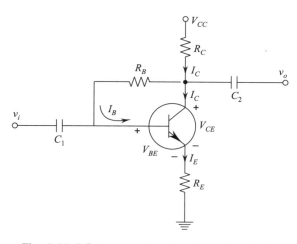

Fig. 3.24 DC bias circuit with voltage feedback.

It may be noted that it is dc analysis, hence, C_1 and C_2 look as though open $(X_C = \dfrac{1}{2\pi f C} = \infty$ for $f = 0)$.

By KVL around the loop from V_{CC} to R_C, R_B, base, emitter, R_E and ground, we get:

$$V_{CC} - I'_C R_C - I_B R_B - V_{BE} - I_E R_E = 0$$

We know that:

$$I'_C = I_C + I_B, \; i.e., \; I'_C = I_C \text{ as } I_B \text{ is very small compared to } I_C$$

and also $\quad I'_C \simeq I_C = \beta I_B$ and $I_E \simeq I_C$.

Hence, the loop equation becomes:

$$V_{CC} - \beta I_B R_C - I_B R_B - V_{BE} - \beta I_B R_E = 0$$

or

$$I_B = \dfrac{V_{CC} - V_{BE}}{R_B + \beta(I_C + R_E)} \qquad \ldots(3.25)$$

$$I_{CQ} = \beta I_B = \dfrac{\beta(V_{CC} - V_{BE})}{R_B + \beta(I_C + R_E)}$$

or

$$I_{CQ} = \dfrac{V_{CC} - V_{BE}}{I_C + R_E} \text{ for } R_B \ll \beta(I_C + R_E) \qquad \ldots(3.26)$$

The collector-emitter loop through R_C and R_E gives:

$$I_E R_E + V_{CE} + I'_C R_C - V_{CC} = 0$$

Taking $I'_C \simeq I_C$ and $I_E \simeq I_C$, we get:

$$I_C(R_C + R_E) + V_{CE} - V_{CC} = 0$$

or

$$V_{CE} = V_{CC} - I_C(R_C + R_E) \qquad \ldots(3.27)$$

3.6.5 Comparison of Biasing Circuits

1. **Fixed Bias Circuit:** The values of collector current I_C and collector emitter voltage V_{CE} dependent upon the value of β which varies with temperature. This implies that operating point Q will change with change in β due to temperature variation. Thus, a stable Q-point cannot be achieved in a fixed-bias circuit. Therefore, it is rarely used.

2. **Collector to Base Bias:** Collector to base biasing has greater stability than fixed bias circuit.

3. **Voltage Divider Bias:** The voltage divider or salt bias circuit gives stability in operating point Q as it is almost independent of β value. Further, even same type of transistor can have different values of β and this also does not affect the stability of point Q.

4. **Emitter Bias Circuit:** Similar to voltage divider bias, this circuit also provides almost same stability of operating point Q.

3.7 GRAPHICAL ANALYSIS OF CE AMPLIFIER

This method requires output characteristics of the transistor which are supplied by the manufacturer. Application of the ac voltage to the input gives variations in base current. The corresponding collector current and collector voltage variation can be seen on the characteristics. The graphical method does not involve any approximations, therefore, the results obtained are more accurate than the equivalent circuit method. The maximum ac voltage which can be properly handled by the amplifier, can also be visualised. Graphical method is the only suitable method for large signal amplifiers, *i.e.*, power amplifiers.

In order to understand this method, a common emitter (*CE*) amplifier circuit of Fig. 3.25 is considered.

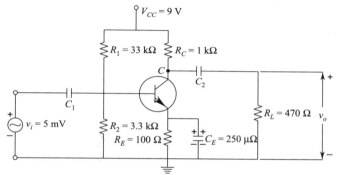

Fig. 3.25 A common-emitter amplifier circuit.

The dc load line (circuit is shown in figure) equation is given by:

$$I_C = \frac{V_{CC} - V_{CE}}{R_C + R_E} = \frac{V_{CC} - V_{CE}}{R_{dc}}, \quad R_{dc} = 1 \text{ k}\Omega \text{ to } 0.1 \text{ k}\Omega = 1.1 \text{ k}\Omega$$

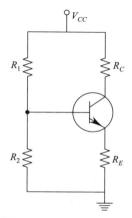

Fig. 3.26 CE amplifier circuit in dc condition.

For, $V_{CE} = 0$,

$$I_C = \frac{V_{CC}}{R_{dc}} = \frac{9\text{ V}}{1.1\text{ k}\Omega} = 8.2\text{ mA}$$

For $I_C = 0$, we get:

$$0 = \frac{V_{CC} - V_{CE}}{R_{dc}} \text{ or } V_{CE} = V_{CC} = 9\text{ V}$$

The dc load line has a slope of $\left(-\dfrac{1}{R_{dc}}\right)$ and is drawn using points (9 V, 0) and (0, 8.2 mA) on the transfer output characteristics in Fig. 3.27. The operating (quiescent) point Q is intersection of dc load-line and the output characteristics of $I_B = 30$ mA. The Q point has the values $I_C = 4$ mA and $V_{CC} = 4.5$ V.

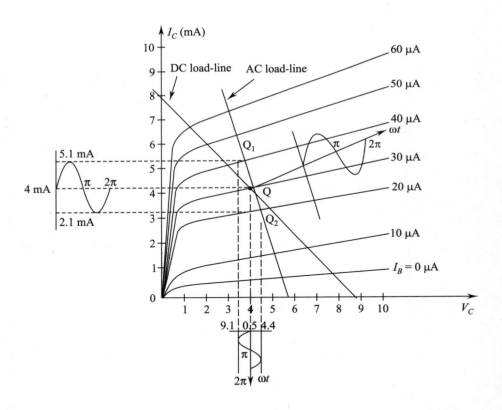

Fig. 3.27 Analysis by graphical method.

The circuit of Fig. 3.28 is applicable in the case of ac input signal V_i. In this circuit, the load resistance in R_C in parallel with R_L. The ac load resistance is given by:

$$R_{ac} = 1 \text{ k}\Omega \parallel 470 \text{ }\Omega = 320 \text{ }\Omega$$

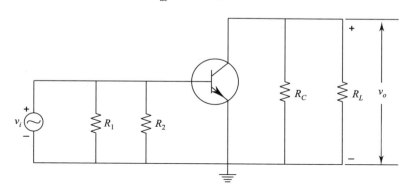

Fig. 3.28 CE amplifier circuit in ac condition.

The ac load line has a slope of $-\dfrac{1}{R_{ac}}$ and passes through point Q as drawn on the output characteristics. The ac input signal is 5 mV, i.e., $5 \times \sqrt{2} \times 2 = 14.14$ mV peak to peak. Now, consider that the input characteristics of the transistor produces a 20 mA peak-to-peak variation in the base current corresponding to the given ac voltage input. 20 mA to 40 mA variation in the base current gives upper and lower operating points Q_1 and Q_2. This gives collector current variation from 2.9 to mA to 5.1 mA. The collector to emitter (V_{CE}) voltage variation is between 4.1 V to 4.9 V. The current gain and voltage gain are calculated as:

Current gain, $\qquad A_i = \dfrac{I_{C\,\max} - I_{C\,\min}}{I_{B\,\max} - I_{B\,\min}} \qquad \qquad \ldots(3.28)$

or $\qquad A_i = \dfrac{(5.1 - 2.9) \text{ mA}}{(40 - 20) \text{ }\mu\text{A}} = 110$

and voltage gain, $\qquad A_v = \dfrac{V_{CE\,\max} - V_{CE\,\min}}{V_{i\,\max} - V_{i\,\min}} \qquad \qquad \ldots(3.29)$

$\qquad\qquad\qquad = \dfrac{(4.9 - 4.1) \text{ V}}{14.14 \text{ mV}}$

or $\qquad A_V = 56.58.$

The input voltage current are in phase whereas collector current and collector-to-emitter voltage are out of phase by 180° as shown in Fig. 3.29.

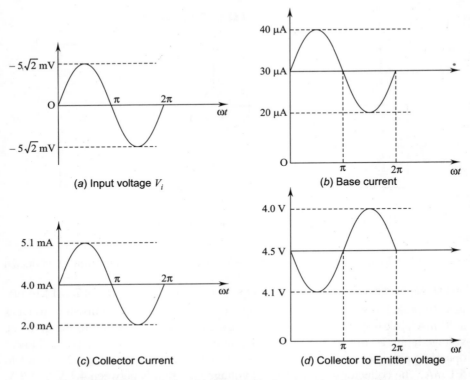

Fig. 3.29 Phase relationships between input and output voltages.

3.8 PARAMETER MODEL

h (hybrid) parameters are mixture of constants having different units. A transistor is a three terminal device which can be represented as two-port network as shown in Fig. 3.30. The left side terminals (1 – 1′) are input and the right side terminals (2 – 2′) are output. The two-part network voltage and current relationships in terms of h-parameter are given by following two equations:

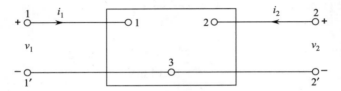

Fig. 3.30 Two port network.

$$v_1 = h_{11} i_1 + h_{12} v_2 \qquad \ldots(3.30)$$
$$i_2 = h_{21} i_1 + h_{22} v_2 \qquad \ldots(3.31)$$

where

$$h_{11} = \left.\frac{v_1}{i_1}\right|_{v_2=0} = \text{input impedance} = h_i$$

$$h_{21} = \left.\frac{i_2}{i_1}\right|_{v_2=0} = \text{forward current ratio} = h_f$$

$$h_{12} = \left.\frac{v_1}{v_2}\right|_{v_2=0} = \text{reverse voltage ratio} = h_r$$

$$h_{22} = \left.\frac{i_2}{v_2}\right|_{i_1=0} = \text{output admittance} = h_o$$

The h-parameter model of a transistor is shown in Fig. 3.31.

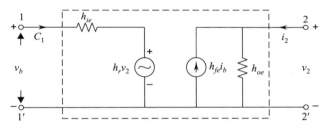

Fig. 3.31 h-parameter model of a transistor.

3.8.1 *h*-Parameter Model of *CE* Amplifier Configuration

An additional suffix is added to the symbols of the h-parameters to indicate that the transistor is used in the CE mode. Hence, the terminal 1 is the base terminal, terminal 2 is the collector and terminals 1' and 2' combined is the emitter. Accordingly, v_1 and i_1 become v_b and i_b; v_2 and i_2 become v_c and i_c. Thus, h-parameter model of the transistor in *CE* mode becomes as shown in Fig. 3.32.

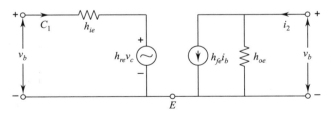

Fig. 3.32 h-parameter model of CE amplifier configuration

Then h-parameter based circuit equation becomes:

$$v_b = h_{ie} i_b + h_{re} v_c \qquad \ldots(3.32)$$
$$i_c = h_{fe} i_b + h_{oe} v_c \qquad \ldots(3.33)$$

Fig. 3.33 shows the complete h-parameter equivalent circuit of a transistor amplifier of CE configuration. It has small signal voltage source having low frequency. If the signal is small, the active region is linear; and if the signal has low frequency, the capacitive effect is negligible. Thus, in such case the h-parameters remain constant.

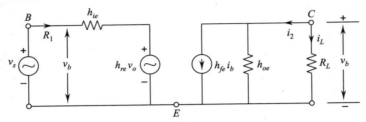

Fig. 3.33 Complete h-parameter model of CE configuration.

3.8.1.1 *Current gain* A_i

The **current gain** of the circuit is given by:

$$A_i = \frac{i_L}{i_b} = -\frac{i_c}{i_b} \text{ as } i_L = -i_c$$

The voltage across the load is: $v_c = i_L R_L = -i_c R_L$

By substituting v_c in equation (3.33), we get:

$$i_c = h_{fe} i_b - h_{oe} i_c R_L$$

or $\qquad i_c (1 + h_{oe} R_L) = h_{fe} i_b$

or

$$\frac{i_c}{i_b} = \frac{h_{fe}}{1 + h_{oe} R_L}$$

∴ Current gain, $\qquad \boxed{A_i = -\frac{h_{fe}}{1 + h_{oe} R_L}} \qquad$ ...(3.34)

3.8.1.2 *Input resistance*

The **input resistance,** $\qquad \boxed{R_i = \frac{v_c}{i_b}}$

By substituting value of $v_c = -i_c R_L$ in equation (3.32), we get:

$$v_b = h_{ie} i_b + h_{re} (-i_c R_L)$$

or $\qquad v_b = h_{ie} i_b - h_{re} i_c R_L$

or $\qquad \frac{v_b}{i_b} = h_{ie} - h_{re} \left(\frac{i_c}{i_b}\right) R_L$

Bipolar Junction Transistor (BJT)

or
$$R_i = \frac{v_b}{i_b} = h_{ie} - h_{re}\left(\frac{i_c}{i_b}\right)R_L$$

$$R_i = h_{ie} - h_{re} \times A_i \times R_L$$

Now, by substituting value of A_i from equation (3.34), we get:

$$R_i = h_{ie} - h_{re} \times \left(-\frac{h_{fe}}{1+h_{oe}R_L}\right) \times R_L$$

or
$$R_i = h_{ie} - \frac{h_{re} \cdot h_{fe}}{h_{oe} + \frac{1}{R_L}} \qquad \ldots(3.35)$$

3.8.1.3 Voltage gain A_v

The **voltage gain** is defined as: $\boxed{A_v = \frac{v_c}{v_i} = -\frac{i_c R_L}{v_i}}$

as
$$A_i = -\frac{i_c}{i_b} \text{ or } i_c = -A_i i_b$$

$$\therefore \quad A_v = \frac{A_i i_b R_L}{v_i} = A_i R_L \left(\frac{i_b}{v_b}\right) \qquad \ldots(3.36)$$

We know that:
$$\frac{v_b}{i_b} = R_i$$

$$\therefore \quad A_v = \frac{A_i R_L}{R_i}$$

By substituting,
$$A_i = -\frac{h_{fe}}{1+h_{oe}R_L}$$

and
$$R_i = h_{ie} - \frac{h_{re} \cdot h_{fe}}{h_{oe} + \frac{1}{R_L}} = h_{ie} - \frac{h_{re} \cdot h_{fe} R_L}{1+h_{oe}R_L}$$

We get
$$A_v = -\frac{h_{fe}}{1+h_{oe}R_L} \times \frac{1}{h_{oe} - \frac{h_{re} h_{fe} R_L}{1+h_{oe}R_L}}$$

$$= -\frac{h_{fe}}{1+h_{oe}R_L} \times \frac{1+h_{oe}R_L}{h_{ie} + h_{ie} h_{oe} R_L - h_{re} h_{fe} R_L}$$

$$= -\frac{h_{fe}}{h_{ie} + \Delta h R_L} \qquad \ldots(3.37)$$

where
$$\Delta h = h_{ie} h_{oe} - h_{re} h_{fe}$$

3.8.1.4 Output resistance

The output resistance can be calculated by opening load R_L and making the circuit signal as zero. Thus, the circuit of Fig. 3.33 becomes as shown in Fig. 3.34.

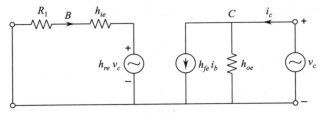

Fig. 3.34 Circuit for calculation of R_o.

Output resistance, $\boxed{R_o = \dfrac{v_c}{i_c}}$

From equation we get: $i_c = h_{fe} i_b + h_{oe} v_c$

$$\therefore \quad R_o = \dfrac{v_c}{h_{fe} i_b + h_{oe} v_c} \qquad ...(3.38)$$

By writing KVL for input side of the circuit, we get:

$$R_s i_b + h_{re} v_c + h_{ie} i_b = 0$$

or $$i_b = -\dfrac{h_{re} v_c}{R_s + h_{ie}} \qquad ...(3.39)$$

Now, by substituting i_b value from equation (3.39) into equation (3.38), we get:

$$R_o = \dfrac{v_c}{h_{fe}\left(\dfrac{-h_{re} v_c}{R_s + h_{ie}}\right) + h_{oe} v_c}$$

or $$R_o = \dfrac{R_s + h_{ie}}{h_{oe}(R_s + h_{ie}) - h_{fe} h_{re}}$$

or $$R_o = \dfrac{R_s + h_{ie}}{R_s h_{oe} + (h_{ie} h_{oe} - h_{fe} h_{re})}$$

or $$R_o = \dfrac{R_s + h_{ie}}{R_s h_{oe} + \Delta h} \quad \text{for } \Delta h = h_{ie} h_{oe} - h_{fe} h_{re}$$

or $$R_o = \dfrac{h_{ie}}{\Delta h} \qquad ...(3.40)$$

for input source resistance, $R_s = 0$

Bipolar Junction Transistor (BJT)

3.9 HYBRID EQUIVALENT CIRCUIT FOR COMMON BASE (CB)

3.9.1 Configuration

In the h-parameter addition subscript shall be added in Fig. 3.35 and the circuit becomes as shown in Fig. 3.34. The h-parameter equations shall be given by:

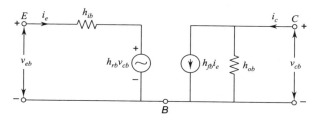

Fig. 3.35 CB circuit with h-parameters.

$$v_{eb} = h_{ib}\, i_e + h_{ob}\, v_{cb} \qquad \ldots(3.41)$$

$$i_c = h_{fb}\, i_e + v_{cb}\, h_{ob} \qquad \ldots(3.42)$$

The current gain, input resistance, voltage gain and output resistance of the CB circuit can be derived similar to CE circuit or it can be obtained from CE formula by replacing e with b in place of additional subscript.

$$A_i = -\frac{h_{fb}}{1 + h_{ob}\, R_L} \qquad \ldots(3.43)$$

$$R_i = h_{ib} - \frac{h_{rb}\, h_{fb}}{h_{ob} + \dfrac{1}{R_L}} \qquad \ldots(3.44)$$

$$A_v = -\frac{h_{fb}\, R_L}{h_{ib} + \Delta h\, R_L} \qquad \ldots(3.45)$$

and

$$R_o = \frac{R_s + h_{ib}}{R_s\, h_{ob} + \Delta h} \qquad \ldots(3.46)$$

where

$$\Delta h = h_{ib}\, h_{ob} - h_{rb}\, h_{fb}$$

3.10 HYBRID EQUIVALENT CIRCUIT FOR COMMON COLLECTOR (CC)

The hybrid equivalent circuit is shown in Fig. 3.36. In this case, the additional subscript shall be c.

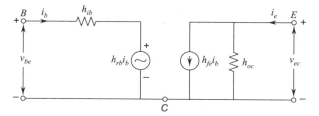

Fig. 3.36 CC circuit with h-parameters.

In this case additional subscript shall be changed to c for current gain, input resistance, voltage gain and output resistance formulae

$$A_i = -\frac{h_{fc}}{1 + h_{oc} R_L} \quad \ldots(3.47)$$

$$R_i = h_{ic} - \frac{h_{rc} h_{fc}}{h_{oc} + \frac{1}{R_L}} \quad \ldots(3.48)$$

$$A_v = -\frac{h_{fc} R_L}{h_{ic} + \Delta h R_L} \quad \ldots(3.49)$$

$$R_o = \frac{R_s + h_{ic}}{R_s h_{oc} + \Delta h} \quad \ldots(3.50)$$

where $\Delta h = h_{ic} h_{oc}$ and $h_{rc} h_{fc}$

3.11 OVERALL CURRENT GAIN

In order to calculate current including source resistance, the source voltage is considered as source current and the circuit is as per Fig. 3.37.

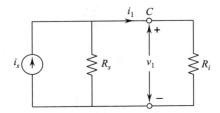

Fig. 3.37 Input circuit with current source and source resistance.

The current gain with source resistance,

$$A_{is} = \frac{i_L}{i_s} = \frac{-i_2}{i_s} = \frac{-i_2}{i_1} \times \frac{i_1}{i_s} = -A_i \times \frac{i_1}{i_s} \text{ as } \frac{i_2}{i_1} = A_i$$

From the circuit 3.37, we get:

$$i_1 = (R_s \| R_i) \times i_s \times \frac{1}{R_i}$$

or

$$i_1 = \frac{i_s}{R_i} \times \frac{R_s \times R_i}{R_s + R_i}$$

or

$$\frac{i_1}{i_s} = \frac{R_s}{R_s + R_i}$$

$$\therefore \quad A_{is} = -A_i \times \frac{R_s}{R_s + R_i} \quad \ldots(3.51)$$

3.12 OVERALL VOLTAGE GAIN

The input part will contain source resistance and the circuit becomes as shown in Fig. 3.38.

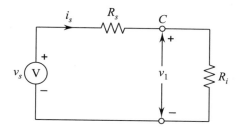

Fig. 3.38 Input circuit with voltages source and source resistance.

The voltage gain with source resistance,

or
$$A_{vs} = A_v \times \frac{v_b}{v_s} \text{ as } \frac{v_c}{v_b} = A_v$$

From the circuit 3.38, we get:

$$v_b = \frac{v_s}{R_s + R_i} \times R_i \text{ or } \frac{v_b}{v_s} = \frac{R_i}{R_s + R_i}$$

$$\therefore \quad A_{vs} = A_v \times \frac{R_i}{R_i + R_s} \quad \ldots(3.52)$$

SOLVED EXAMPLES

Example 3.1 A transistor is connected in CB configuration. When the emitter voltage is changed by 200 mV, the emitter current changes by 5 mA. During this variation, the collector to base voltage is kept fixed. Calculate the dynamic input resistance of transistor.

Solution: The dynamic input resistance of transistor,

$$r_i = \frac{\Delta v_{EB}}{\Delta i_E}\bigg|_{V_{CB} = \text{constant}}$$

or
$$r_i = \frac{200 \text{ mV}}{5 \text{ mA}}$$

or
$$r_i = 40 \text{ }\Omega.$$

Example 3.2 The figure given shows the collector-base bias circuit with $\beta = 100$. Assuming $V_{BE} = 0$, determine the following:
 (i) the value of I_B
 (ii) the value of I_C
 (iii) the value of V_{CE}
 (iv) the stability factor

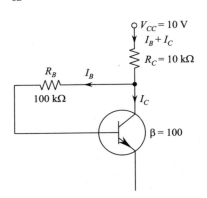

Solution:

(i) The value of base current, $I_B = \dfrac{V_{CC} - V_{BE}}{R_B + \beta R_C} = \dfrac{(10-0)\,V}{(100+100\times 10)\,k\Omega}$

or $I_B = 0.09 \times 10^{-4}$ mA.

(ii) The value of collector current, $I_C = \beta I_B = 100 \times 9$ mA

or $I_C = 0.9$ mA.

(iii) The value of voltage from collector to emitter,
$$V_{CE} = V_{CC} - I_C R_C$$
$$= (10 - 0.9 \times 10^{-3} \times 10 \times 10^3)\,V$$

or $V_{CE} = 1$ V.

(iv) The stability factor, $S = \dfrac{1+\beta}{1+\beta\left(\dfrac{R_C}{R_C + R_B}\right)}$

or $S = \dfrac{1+100}{1+100\left(\dfrac{10}{10+100}\right)}$

or $S = \dfrac{101}{1+0.09} = \dfrac{101}{1.09}$

or $S = 92.6$.

Example 3.3 A transistor with $\beta = 100$ is used in CE configuration. The collector circuit resistance is $R_C = 1$ kΩ and $V_{CC} = 20$ V. Assuming $V_{BE} = 0$, find the value of collector to base resistance, such that quiescent collector-emitter voltage, is 4 V. Also determine the stability factor in this case.

Solution: Consider the figure shown:

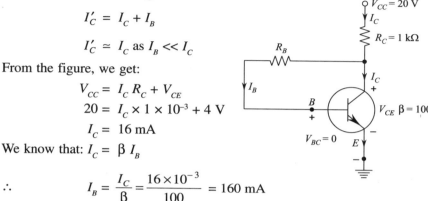

$$I'_C = I_C + I_B$$

or $$I'_C \simeq I_C \text{ as } I_B \ll I_C$$

From the figure, we get:

$$V_{CC} = I_C R_C + V_{CE}$$

or $$20 = I_C \times 1 \times 10^{-3} + 4 \text{ V}$$

or $$I_C = 16 \text{ mA}$$

We know that: $I_C = \beta I_B$

$\therefore$ $$I_B = \frac{I_C}{\beta} = \frac{16 \times 10^{-3}}{100} = 160 \text{ mA}$$

From figure, it is seen:

$$V_{CE} = I_B R_B + V_{BE}$$

or $$V_{CE} = I_B R_B + 0$$

or $$R_B = \frac{V_{CE}}{I_B} = \frac{4}{160 \times 10^{-6}}$$

or $$R_B = 25 \text{ k}\Omega.$$

The stability factor,

$$S = \frac{\beta + 1}{1 + \dfrac{\beta R_C}{R_B + R_C}} = \frac{100 + 1}{1 + \dfrac{100 \times 1}{25 + 1}} = \frac{101}{1 + \dfrac{100}{26}}$$

or $$S = \frac{101}{4.84} = 20.86.$$

Example 3.4 In common emitter circuit as given below an NPN transistor having a value of $\beta = 50$ is used with $V_{CC} = 10$ V and $R_C = 2$ kΩ. If a 100 kΩ resistor is connected between collector and base and $V_{BE} = 0$, determine:

(i) the position of quiescent point and (ii) stability factor, S.

Solution:

(i) Consider the figure shown:

$$I'_C = I_C + I_B$$

or $$I'_C \simeq I_C \text{ as } I_B \ll I_C$$

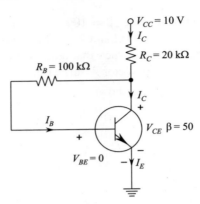

KVL from V_{CC}, collector, R_B, base, emitter and ground gives:

$$V_{CC} = I'_C R_C + I_B R_B + V_{BE}$$

or $$V_{CC} = I_C \times R_C + \frac{I_C}{\beta} \times R_B + 0 \text{ as } I'_C \approx I_C \text{ and } I_C = b\, I_B$$

or $$V_{CC} = I_C \left[R_C + \frac{R_B}{\beta} \right]$$

or $$I_C = \frac{V_{CC}}{R_C + \frac{R_B}{\beta}} = \frac{10 \text{ V}}{\left(2 + \frac{100}{50}\right) \text{k}\Omega}$$

or $$I_C = 2.5 \text{ mA}$$

We know: $$I_C = \beta\, I_B$$

∴ $$I_B = \frac{I_C}{\beta} = \frac{2.5 \text{ mA}}{50} = 0.05 \text{ mA}$$

We can see that: $V_{CE} = I_B R_B + V_{BE}$

or $V_{CE} = 0.05 \times 10^{-3} \times 100 \times 10^3 + 0$

or $V_{CE} = 5 \text{ V}$

∴ Point, $Q = (5 \text{ V}, 2.5 \text{ mA})$.

The stability factor, $$S = \frac{\beta + 1}{1 + \frac{\beta R_C}{R_B + R_C}}$$

or $$S = \frac{50 + 1}{1 + \frac{50 \times 20}{100 + 20}} = \frac{51 \times 102}{102 + 100}$$

or $S = 25.75$.

Example 3.5 In a NPN transistor amplifier stage, a 9 V battery, supply is to be used with collector to base biasing. Determine the values of R_B and R_C if I_{CBO} is negligible, $\beta = 100$ and if the quiescent point is specified by $I_C = 0.2$ A and $V_{CE} = 5$ V.

Solution: Consider the circuit given in the figure.

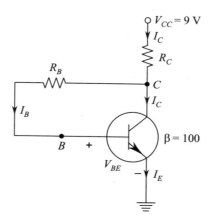

From the figure, we get:

$$V_{CC} = I'_C R_C + V_{CE}$$

or
$$V_{CC} = (I_C + I_B) R_C + V_{CC}$$

or
$$V_{CC} = I_C R_C + V_{CE} \text{ as } I_B \ll I_C$$

∴
$$I_C R_C = V_{CC} - V_{CE} = 9 - 5 = 4 \text{ V}$$

as
$$V_{CC} = 9 \text{ V} \quad \text{and} \quad V_{CE} = 5 \text{ V}$$

∴
$$R_C = \frac{4}{I_C} = \frac{4}{0.2} = 20 \text{ }\Omega.$$

It is known that:
$$I_C = \beta I_B$$

or
$$I_B = \frac{I_C}{\beta} = \frac{0.2}{100} = 2 \text{ mA}$$

From the figure, we get: $V'_{CE} = I_B R_B + V_{BE}$

or
$$5 \text{ V} = 2 \text{ mA} \times R_B \text{ taking } V_{BE} = 0$$

or
$$R_B = \frac{5 \text{ V}}{2 \text{ mA}} = 2.5 \text{ k}\Omega.$$

Example 3.6 In the given figure a transistor with $\beta = 45$ is used with collector to base resistor (R) biasing, with a quiescent value of 5 V for V_{CE}. If $V_{CC} = 24$ V, $R_L = 10$ kW and $R_E = 270$ W, find the value of (*i*) R, and (*ii*) stability factor.

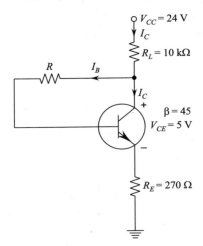

Solution: Consider circuit of the figure given here

(*i*) By KVL in collector and emitter circuit, we get:

$$V_{CC} = I'_C R_L + V_{CE} + I_E R_E$$

or $$V_{CC} = (I_C + I_B) R_L + V_{CE} + (I_B + I_C) R_E$$

as $$I'_C = I_C + I_B \quad \text{and} \quad I_E = I_B + I_C$$

or $$V_{CC} = (\beta I_B + I_B) R_L + V_{CE} + (\beta I_B + I_B) R_E$$

or $$V_{CC} = (\beta I_B + I_B)(R_L + R_E) + V_{CE}$$

or $$24 \text{ V} = (45 + 1) \times I_B \times (10 \text{ k}\Omega + 0.2070 \text{ k}\Omega) + 5 \text{ V}$$

or $$I_B = \frac{19 \text{ V}}{46 \times (10 + 0.27) \times 10^3 \, \Omega}$$

or $$I_B = \frac{19}{46 \times 10.27} \text{ mA}$$

or $$I_B = 40 \text{ mA}.$$

We can see from the circuit of the figure that:

$$V_{CE} = V_{BE} + R I_B$$

or $$R = \frac{V_{CE} - V_{BE}}{I_B} = \frac{5 - 0.6}{40 \times 10^{-6}} \Omega$$

where $V_{BE} = 0.6$ is taken.

or $$R = 110 \text{ k}\Omega.$$

Bipolar Junction Transistor (BJT)

(ii) Stability factor, $S = \dfrac{\beta + 1}{1 + \dfrac{\beta R_E}{R_E + R}}$

$= \dfrac{45 + 1}{1 + \dfrac{45 \times 270}{270 + 110 \times 10^3}}$

$= \dfrac{46}{1 + \left(\dfrac{12150}{110.270 \times 10^3}\right)}$

$= \dfrac{46}{1 + 0.11} = \dfrac{46}{1.11}$

or $\qquad S = 41.44.$

Example 3.7 In a CE germanium transistor amplifier, self-bias is used. The various parameters are: $V_{CC} = 16$ volts, $R_C = 3$ kΩ, $R_E = 2$ kΩ, $R_1 = 56$ kΩ, $R_2 = 20$ kΩ and $\alpha = 0.985$. Determine the following:

(i) Operating point

(ii) the stability factor, S

Solution: Consider the circuit of the figure being given here.

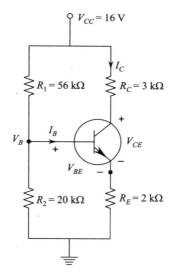

$\beta = \dfrac{\alpha}{1-\alpha} = \dfrac{0.985}{1 - 0.985}$ as $\alpha = 0.985.$

(i) or $\beta = 66$ from the circuit, $V_B = \dfrac{V_{CC} \times R_2}{R_1 + R_2} = \dfrac{16 \times 20 \times 10^3}{(56 + 20) \times 10^3} = 4.21$ V.

$$I_C = \frac{V_B - V_{BE}}{R_E}$$

or $$I_C = \frac{(4.21 - 0.3)\ V}{2 \times 10^3\ \Omega} \text{ as } V_{BE} = 0 \text{ for germanium transistor}$$

By application of KVL in the collector-emitter loop, we get:
$$V_{CE} = V_{CC} - I_C(R_C + R_E)$$
$$= 16\ V - 2.0 \times 10^{-3}(3 + 2) \times 10^3 = 6\ V$$

Thus, operating point, $Q = (6\ V, 2\ mA)$.

(*ii*) Stability factor, $$S = \frac{1 + \beta\left(1 + \dfrac{R_{TH}}{R_E}\right)}{1 + \beta + \dfrac{R_{TH}}{R_E}}$$

where $$R_{TH} = R_1 \| R_2 = \frac{R_1 \cdot R_2}{R_1 + R_2} = \frac{56 \times 10^3 \times 20 \times 10^3}{(56 + 20) \times 10^3} = 14.73\ k\Omega.$$

$$\therefore \quad S = \frac{1 + 66\left(1 + \dfrac{14.73}{2}\right)}{1 + 66 + \dfrac{14.73}{2}}$$

or $\quad S = 7.5$.

Example 3.8 Calculate the collector current and collector to emitter voltage of the circuit figure assuming the following circuit components and transistor specifications.

$R_1 = 40\ k\Omega$, $R_2 = 4\ k\Omega$, $R_C = 10\ k\Omega$, $R_E = 1.5\ k\Omega$,
$V_{BE} = 0.5\ V$, $B = 40$, $V_{CC} = 22\ V$.

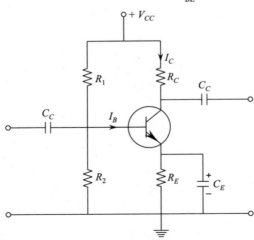

Solution: For given self-bias circuit, the value of collector current I_C is given by

$$I_C = \frac{V_B - V_{BE}}{R_E} \qquad ...(1)$$

But base voltage

$$V_B = V_{CC} \times \frac{R_2}{R_1 + R_2}$$

$$V_B = 22 \times \frac{4 \times 10^3}{(40 + 4) \times 10^3}$$

$$= \frac{22 \times 4}{44} = 2 \text{ Volt}$$

Using equation (i)

$$I_C = \frac{V_B - V_{BE}}{R_{BE}}$$

$$= \frac{2 - 0.5}{1.5 \times 10^3} = \frac{1.5}{1.5 \times 10^3} = 1 \text{ mA}$$

$$I_C = 1 \text{ mA}.$$

Now, collector to emitter voltage

$$V_{CE} = V_{CC} - I_C (R_C + R_E)$$
$$= 22 - 1 \times 10^{-3} (10 + 1.5) \times 10^3$$
$$V_{CE} = 22 - 11.5 = 8.5 \text{ volt}$$
$$I_C = 1 \text{ mA}$$
$$V_{CE} = 8.5 \text{ volt}.$$

Example 3.9 In a single state CE amplifier, $V_{CC} = 20$ volt, $b = 50$, $R_E = 200$ W, $R_1 = 60$ kW and $R_2 = 30$ kW. Determine the dc voltage across R_E.

Solution: The base voltage V_B (the voltage at base w.r.t. ground) is

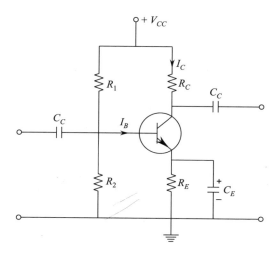

$$V_B = \frac{V_{CC} \times R_2}{R_1 + R_2} \quad [\because V_B \text{ is the voltage across } R_2]$$

$$= 20 \times \frac{30 \times 10^3}{(60+30) \times 10^3}$$

$$= \frac{20 \times 30}{90}$$

$$V_B = \frac{20}{3} = 6.67 \text{ volt.}$$

The voltage across the emitter resistor R_E is given by

$$V_E = V_B - V_{BE}$$

Taking $V_{BE} \cong 0.6$ volt

$$V_E = 7.66 - 0.6$$

$$V_E = 7.07 \text{ volt.}$$

Example 3.10 A silicon transistor with $V_{BE} = 0.8$, $h_{FE} = 100$, V_{CE} sat $= 0.2$ V, is used in the circuit shown in the given figure. Find the minimum value of R_C for which the transistor reaches its saturation.

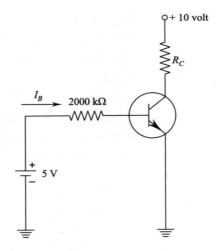

Solution: Base current I_B is given by

$$I_B = \frac{5 \text{ V}}{200 \text{ k}\Omega} = \frac{5}{200 \times 10^3}$$

$$I_B = \frac{5}{2} \times 10^{-5} = 25 \text{ mA}$$

Collector current $\quad I_C = \beta\, I_B = 100 \times 25 \times 10^{-6}$

$$= 25 \times 10^{-4} = 2.5$$

From the figure, $V_{CC} = V_{CE} + I_C R_C$
$$10 = 0.2 + 2.5 \times 10^{-3} \times R_{C\,(min)}$$
$$9.8 = 2.5 \times 10^{-3} \times R_{C\,(min)}$$

$$R_{C\,(min)} = \frac{9.8}{2.5}\ k\Omega = 3.92\ k\Omega.$$

Example 3.11 For the circuit shown in given figure assume $h_{FE} = 100$ and $V_{BE} = 0.8$ V.

(a) Find if the silicon transistor is in cut-off, saturation or active region.
(b) Find V_C.
(c) Find the minimum value of the Emitter Resistance for which the transistor operates in active region.

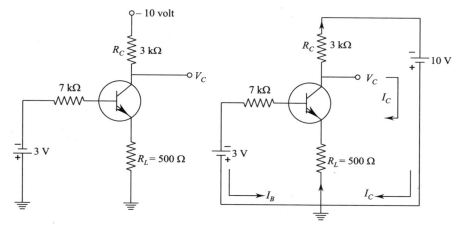

Solution: R_C drawing the given circuit
Applying KVL to the input loop
$$7 \times 10^3\ I_B + 0.8 + 500\ (I_B + I_C) = 3$$
or
$$7500\ I_B + 500\ I_C = 2.2 \qquad \ldots(i)$$
Applying KVL to the output loop
$$3 \times 10^3\ I_C + V_{CE} + 500\ (I_B + I_C) = 10$$
Taking limiting value of $V_{CE} = 0.2$ for saturation as in last question
$$3000\ I_C + 0.2 + 500\ I_B + 500\ I_C = 10$$
or
$$500\ I_B + 3500\ I_C = 9.8 \qquad \ldots(ii)$$
Solving (i) and (ii) we have
$$I_C = 2.78\ mA,\ I_B = 0.1\ mA$$
For saturatioin,
$$I_{B\,(min)} = \frac{I_C}{h_{FE}} = \frac{2.78\ mA}{100}$$
$$I_{B\,(min)} = 0.0278\ mA.$$

But, $I_B = 0.1$ mA, which is greater than 0.0278 mA.
$$I_B = 0.1 \text{ mA} > I_{B \text{ (min)}} = 0.0278 \text{ mA}.$$
Therefore, the transistor is in saturation region.

(b) Since, transistor is in saturation,
$$V_C = V_{BE} = 0.8 \text{ volt}$$
From circuit
$$V_{CE} = V_C - V_E$$
$$= V_C - 500 (I_C + I_B)$$
$$= 0.8 - 500 (2.78 \times 10^{-3} + 0.1 \times 10^{-3})$$
$$V_{CE} = 0.8 - 500 \times 2.88 \times 10^{-3}$$
$$= 0.64 \text{ volt}.$$
Again from circuit, $V_{CE} = V_{CC} - I_C (R_C + R_E)$
$$- 0.64 = 10 - 2.78 \times 10^{-3} (3 \times 10^3 + R_E)$$
Solving above equation, we get
$$R_E = 827 \text{ } \Omega.$$

Example 3.12 A CE amplifier employing an NPN transistor has load resistor R_C connected between collector and V_{CC} supply of + 16 V. For biasing, a resistor R_1 is connected between collector and base, resistor $R_2 = 30$ kΩ is connected between base 8 ground and resistor $R_E = 1$ kΩ is connected between emitter and ground. Draw the circuit diagram. Calculate the values of R_1 and R_C and the stability factor S if $V_{BE} = 0.2$ V, $I_E = 2$ mA, $a_0 = 0.985$ and $V_{CE} = 6$ V.

Solution: Given that emitter current $I_E = 2$ mA

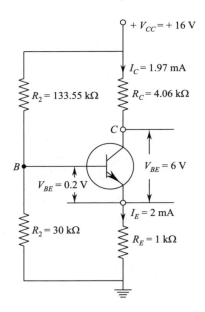

Collector current,
$$I_C = \alpha_0 I_E$$
$$I_C = 0.985 \times 2 = 1.97 \text{ mA}.$$

Also, base current, $I_B = I_E - I_C$
$= 2 - 1.97 = 0.03$ mA

We know that $\beta = \dfrac{\alpha}{1-\alpha}$

Therefore, $\beta = \dfrac{0.985}{1-0.985} = 65.6677$

Collector resistance, $R_C = \dfrac{V_{CC} - V_{CE} - I_E R_E}{I_C}$

$R_C = \dfrac{16 - 6 - 2\text{ mA} \times 1\text{ k}\Omega}{1.97\text{ mA}} = 4.06$ kΩ.

or voltage drop across resistor R_2

$$V_{th} = \dfrac{V_{CC} \times R_2}{R_1 + R_2} = \dfrac{30 \times 16}{30 + R_1} = \dfrac{480}{30 + R_1} \quad \ldots(i)$$

Again we know that $R_{th} = \dfrac{R_1 \cdot R_2}{R_1 + R_2} = \dfrac{30 R_1}{30 + R_1}$

and $V_{th} = I_B R_{th} + V_{BE} + I_E R_E$

or $\dfrac{480}{30 + R_1} = 0.03 \times \dfrac{30 R_1}{30 + R_1} + 0.2 + 2\text{ mA} \times 1\text{ k}\Omega$

or $480 = 0.9 R_1 + 2.2 (20 + R_1)$

Solving, we get $R_1 = 133.55$ kΩ.

For voltage-divider biasing, stability factor S is expressed as

Stability factor, $S = \dfrac{1+\beta}{1 + \dfrac{\beta R_E}{R_{TH} + R_E}}$

Substituting all the values, we get

$$S = \dfrac{1 + 65.667}{1 + 65.667 \dfrac{1}{\dfrac{133.55 \times 30}{133.55 + 30} + 1}} = 18.65.$$

Example 3.13 Assume that a silicon transistor with $\beta_0 = 50$, $V_{BE} = 0.6$ V and $V_{CC} = 20$ V and $R_C = 4.7$ kΩ is used in a self bias circuit. It is designed to establish a Q pt at $V_{CE} = 8$ V and $I_C = 2$ mA and stability factor $S \leq 5.0$. Design the circuit with all component values.

Solution: Given that

Collector current, $I_C = 2$ mA

and $\beta_0 = 50$

Base current, $I_B = \dfrac{I_C}{\beta_0} = \dfrac{2}{50}$

Emitter current is given as

$$I_E = I_B + I_C = 0.04 + 2 = 2.04 \text{ mA}$$
$$V_{CE} = 8 \text{ V}$$
$$V_{BE} = 0.6 \text{ V}, V_{CC} = 20 \text{ V}$$

We know that for a self biased circuit

Emitter resistance $\quad R_E = \dfrac{V_{CC} - V_{CE}}{I_C} - R_C$

Substituting all the values, we get

$$R_E = \dfrac{20 - 8}{2 \text{ mA}} - 4.7 = 1.3 \text{ k}\Omega$$

Also, stability factor for self bias circuit is given as

$$S = (\beta + 1) \dfrac{1 + \dfrac{R_{TH}}{R_E}}{1 + \beta + \dfrac{R_{TH}}{R_E}}$$

Substituting all the values, we get

$$S = (1 + 50) \dfrac{1 + \dfrac{R_{TH}}{1.3}}{1 + 50 + \dfrac{R_{TH}}{1.3}}$$

Solving, we get $\quad R_{TH} = 5.765 \text{ k}\Omega$

Also, $\quad V_{TH} = I_B R_{TH} + V_{BE} + I_E R_E$

or $\quad V_{TH} = 0.04 \times 5.765 + 0.6 + 2.04 \times 1.3$

$\quad\quad = 3.486 \text{ V}$

Again, $\quad V_{TH} = \dfrac{V_{CC} \cdot R_2}{R_1 + R_2} = \dfrac{V_{CC} \cdot R_{TH}}{R_1}$

or $\quad R_1 = R_{TH} \cdot \dfrac{V_{CC}}{V_{TH}}$

Substituting given values, we get

$$R_1 = 5.765 \times \dfrac{20}{3.4826} = 331.3 \text{ k}\Omega.$$

Now since, $\quad V_{TH} = \dfrac{V_{CC} \cdot R_2}{R_1 + R_2}$

Therefore, $\quad R_2 = \dfrac{V_{TH} \times R_1}{V_{CC} - V_{TH}} = \dfrac{3.4826 \times 33.1}{20 - 3.4826}$

$\quad\quad = 7.98 \text{ k}\Omega.$

Bipolar Junction Transistor (BJT)

Example 3.14 The given figure shows a self biased transistor amplifier using a Si transistor with $V_{CC} = 20$ V, $h_{FE} = 400$ and $V_{BE} = 0.65$ V. The transistor should be biased at $V_{CE} = 10$ V and $I_C = 0.6$ mA. Find the value of R_C, R_E, R_1 and R_2 such that it meets the following specification over the temperature range 25°C to 145°C.

$$\frac{\Delta I_C}{\Delta I_C} \leq 10°, V_{BE} \text{ at } 25°C = 650 \pm 50 \text{ mA,}$$

I_{CO} at 25°C = 5 nA max, I_{CO} at 145° = 3.0 mA max

Assume that percentage change in I_C due to V_{BE} and I_{CO} as same (5%).

Solution: Change in I_C due to V_{BE} and I_{CO} is 5%.

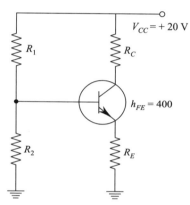

Change in collector current

$$\Delta I_C = \frac{5}{100} \times 0.6 = 0.03 \text{ mA}$$

Change in reverse saturation current

$$\Delta I_{CO} = 3 \times 10^{-6} - 0.005 \times 10^{-6} = 2.995 \text{ mA}$$

Stability factor, $$S = \frac{\Delta I_C}{\Delta I_{CO}}$$

or $$S = \frac{0.03 \times 10^{-3}}{2.995 \times 10^{-6}} = 10$$

For a Si transistor, V_{BE} decreases at the rate of -2.5 mV/°C

Therefore, $$\Delta V_{BE} = \frac{-2.5(145-25)}{-300 \text{ mV}}$$

Now, $$S_V = \frac{\Delta I_C}{\Delta V_{BE}}$$

or $$S_V = \frac{0.03 \times 10^{-3}}{-300 \times 10^{-3}} = -0.0001$$

$$\beta = h_{FE} = 400$$

and
$$S_V = \frac{-S}{R_{TH} + R_E} \times \frac{\beta}{\beta + 1}$$

or
$$-0.0001 = \frac{-10}{R_{TH} + R_E} \times \frac{400}{401}$$

or
$$R_{TH} + R_E = 99{,}750.6$$

Again,
$$S = \frac{R_E + R_{TH}}{R_E + \frac{R_{TH}}{\beta + 1}}$$

or
$$10 = \frac{R_E + R_{TH}}{R_E + \frac{R_{TH}}{\beta + 1}}$$

$$R_E = \frac{391}{3{,}609} R_{TH}$$

Solving equations (*i*) and (*ii*), we get
$$R_E = 9.75 \text{ k}\Omega.$$

and
$$R_{TH} = 90 \text{ k}\Omega$$
$$\Delta I_C = S I_{CO} + S_V \Delta V_{BE} = 0.03 \times 10^{-3} + 0.03 \times 10^{-3} = 0.06 \text{ mA}$$

For the given circuit, we have
$$V_{CC} = I_C R_C + V_{CE} + I_E R_E$$

$$20 = 0.66 R_C + 10 + \left[0.66 + \frac{0.66}{400}\right] \times 9.75$$

or
$$R_C = 5.377 \text{ k}\Omega.$$

Now,
$$V_{TH} = I_B R_{TH} + V_{BE} + I_E R_E$$

Substituting all the values, we get

$$V_{TH} = \frac{0.66}{400} \times 90 + 0.65 + \left[0.66 + \frac{0.66}{400}\right] \times 9.75 = 7.25 \text{ V}$$

Now since
$$R_1 = R_{TH} \frac{V_{CC}}{V_{TH}} = 90 \times \frac{20}{7.25} = 248 \text{ k}\Omega.$$

Also
$$R_2 = \frac{V_{TH} \times R_1}{V_{CC} - V_{TH}}$$

Substituting values, we get
$$R_2 = \frac{7.25 \times 2.48}{20 - 7.25} = 141 \text{ k}\Omega.$$

As R_1, R_2, R_C and R_E are designed for variation up to 145°C, the variation of V_{BE} at 25°C of 650 ± 50 mV will be absorbed in the overall variation.

Example 3.15 Find I_C and V_{CE} for the following circuit. What will happen to V_{CE} if β increases due to temperature?

Solution: Applying KVL to the base emitter loop:

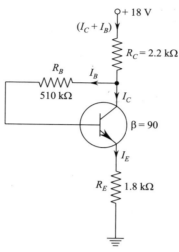

$$V_{CC} = (I_C + I_B) R_C + I_B \times R_B + V_{BE} + I_E R_E$$
$$V_{CC} = I_C R_C + I_B R_B + V_{BE} + I_E R_E$$

∵ $(I_C \gg I_B)$

But $I_E \cong I_C$ and neglecting V_{BE}

We have $V_{CC} = I_C R_C + I_B R_B + 0 + I_C R_E$
$= I_C (R_C + R_E) + I_B R_B$

But $I_C = \beta I_B$

Therefore, $V_{CC} = \beta I_B (R_C + R_E) + I_B R_B$
$= I_B [b (R_C + R_E) + R_B]$

$$I_B = \frac{V_{CE}}{R_B + \beta(R_C + R_E)} = \frac{18}{510 \times 10^3 + 90(2.2 + 1.8) 10^3}$$

$$= \frac{18 \times 10^{-3}}{870} = 0.02 \times 10^{-3}$$

$I_B = 20 \times 10^{-6}$ A = 20 mA

∴ $I_C = b I_B = 90 \times 20 \times 10^{-6}$

Hence $I_C = 1800 \times 10^{-6} = 1.8 \times 10^{-3} = 1.8$ mA

Now, $V_{CE} = I_B R_B + V_{BE}$

Taking $V_{BE} \cong 0$

$V_{CE} \cong I_B R_B = 20 \times 10^{-6} \times 510 \times 10^3 = 20 \times 510 \times 10^{-3} = 102 \times 10^{-2}$

$V_{CE} = 10.2$ volt.

Example 3.16 A transistor has an emitter current of 10 mA and a collector current of 9.95 mA. Calculate its base current.

Solution: Given that $I_E = 10$ mA

and $I_C = 9.95$ mA

Emitter current is given by $I_E = I_C + I_B$

$I_B = I_E - I_C = 10 - 9.95 = 0.05$ mA.

$I_B = 0.05$ mA

Example 3.17 Draw N-P-N and P-N-P transistors. Label all the currents and show the direction of flow. How are all the currents of a transistor related?

Solution: N-P-N Transistor symbol and common emitter circuit.

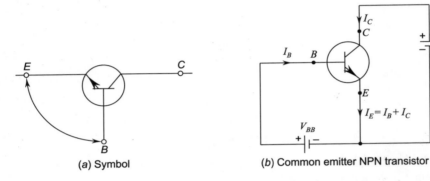

(a) Symbol (b) Common emitter NPN transistor

P-N-P Transistor and common emitter circuit

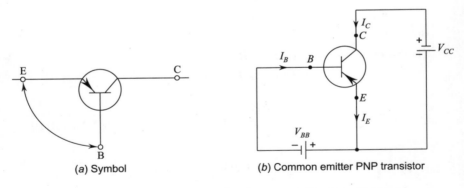

(a) Symbol (b) Common emitter PNP transistor

The currents of a transistor are related by the following relation:

$$I_E = I_B + I_C$$

where I_E = emitter current

I_C = collector current

I_B = base current

Bipolar Junction Transistor (BJT)

Example 3.18 Find the value of β, V_{CC} and R_B in the circuit shown below.

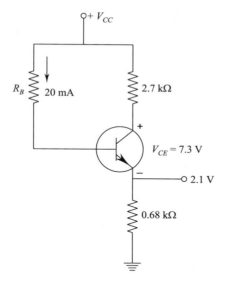

Solution: Given that:

$$I_B = 20 \text{ mA} = 0.02 \text{ mA}$$
$$R_C = 2.7 \text{ k}\Omega$$
$$V_{CE} = 7.3 \text{ V}, V_E = 2.1 \text{ volt}$$
$$R_E = 0.68 \text{ k}\Omega$$
$$V_{BE} = 0.7 \text{ V (assumed)}$$

From the figure, we have

$$I_E R_E = V_E$$

or
$$I_E = \frac{V_E}{R_E}$$

Substituting values, we get

$$I_E = \frac{2.1}{0.68} = 3.09 \text{ mA}$$

Also, since $I_E = I_C + I_B$

Therefore, $I_C = I_E - I_B = 3.09 - 0.02$

or $I_C = 3.07 \text{ mA}$

Further, $I_C = \beta I_B$ so that $\beta = \dfrac{I_C}{I_B} = \dfrac{3.07}{0.02}$,

$$\beta = 154.$$

Applying KVL to the output side of the given circuit, we get
$$V_{CC} = I_C R_C + V_{CE} + V_E$$
Substituting values, we get
$$V_{CC} = 3.07 \times 2.7 + 7.3 + 2.1$$
$$= 1.77 \text{ volt}$$
Also applying KVL to the input side, we obtain
$$V_{CC} = I_B R_B + V_{BE} + V_E$$
or
$$R_B = \frac{V_{CC} - V_{BE} - V_E}{I_B}$$
Putting values, we get
$$R_B = \frac{17.7 - 0.7 - 2.1}{0.02 \times 10^{-3}}$$
$$R_B = 745 \text{ k}\Omega.$$

Example 3.19 Consider a d.c. bias circuit with voltage feedback as in the given figure. Determine the quiescent levels of I_{CQ} and V_{CEQ}. The β of the transistor 1590, and cut in voltage is 0.7 V.

Solution:

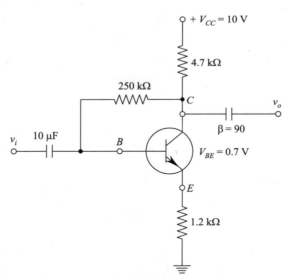

Now, applying KVL to the input side, we get
$$V_{CC} = (I_C + I_B) R_C + I_B R_B + V_{BE} + I_E R_E$$
or
$$V_{CC} = (I_C + I_B) R_C + I_B R_B + V_{BE} + (I_C + I_B) R_E$$
$$V_{CC} = (I_C + I_B)(R_C + R_E) + I_B R_B + V_{BE}$$
But
$$I_B = \frac{I_C}{\beta}$$

Therefore, we have
$$V_{CC} = \left[I_C + \frac{I_C}{\beta}\right](R_C + R_B) + \frac{I_C}{\beta} \cdot R_C + V_{BE}$$

or
$$V_{CC} = \left(\frac{I_C}{\beta}\right)(R_C + R_E) + \frac{I_C}{\beta} \cdot R_B + V_{BE}$$

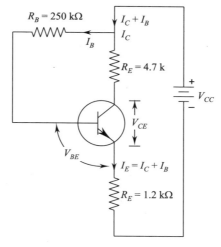

The d.c. equivalent circuit for the given circuit.

or
$$V_{CC} - V_{BE} = I_C\left[\left(1 + \frac{1}{\beta}\right)(R_C + R_E) + \frac{R_B}{\beta}\right]$$

Then, we have
$$I_{CQ} = \frac{\beta(V_{CC} - V_{BE})}{(1+\beta)(R_C + R_E) + R_B}$$

Given
$\beta = 90°$, $V_{CC} = 10$ V, $V_{BE} = 0.7$ V
$R_C = 4.7$ K $= 4700$ ohm, $R_B = 250$ K $= 25000$ ohm

Substituting all these values in equation (i), we get

$$I_{CQ} = \frac{90(10 - 0.7)}{(1+90)(4700 + 100) + 25000}$$

$$I_{CQ} = \frac{90 \times 9.3}{91 \times 5900 + 25000} = 1.06 \text{ mA}.$$

Further, applying KVL to the output side, we get

$$V_{CEQ} = V_{CC} - I_C + \frac{I_C}{\beta}(R_C + R_E)$$

Again, substituting given values, we obtain

$$V_{CEQ} = 3.68 \text{ volts}.$$

Example 3.21 For the emitter bias circuit shown in the given figure, determine I_B, I_C, V_{CE}, V_C, V_B and V_C.

Solution: Note that the dc equivalent circuit is obtained by making the capacitors open circuited.

Applying KVL to the input side, we get

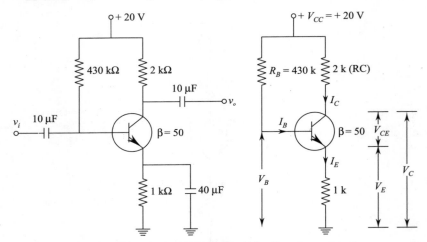

The dc equivalent circuit for the given circuit.

$$V_{CC} = I_B R_B + V_{BE} + I_E R_E \qquad \ldots(i)$$
$$I_E = I_B + I_C \quad \text{and} \quad I_C = \beta I_B$$

With above two substitutions equation (i) becomes

$$V_{CC} = I_B R_B + V_{BE} + (I_B + \beta I_B) R_E$$

Simplifying, we get

$$I_B = \frac{V_{CC} - V_{BE}}{R_B + (\beta + 1) R_E}$$

$V_{BE} \cong 0.7$ volts for Si transistor

Substituting all the given values, we obtain

$$I_B = \frac{20 - 0.7}{430 \times 10^3 + (50 + 1) \times 1 \times 10^3}$$

or $\qquad I_B = 4 \times 10^{-5}$ amp.

Also $\qquad I_C = \beta I_B = 50 \times 4 \times 10^{-5}$
$\qquad I_C = 2$ mA.

Now, applying KVL to the output side, we get

$$V_{CC} = I_C R_C + V_{CE} + I_E R_E$$
or $\qquad V_{CC} = I_C R_C + V_{CE} + (I_C + I_B) R_E$
or $\qquad V_{CE} = V_{CC} - I_C R_C - (I_C + I_B) R_E$
or $\qquad V_{CE} = 20 - 2 \times 10^{-3} \times 2 \times 10^3 - (2 \times 10^{-3} + 4 \times 10^{-6}) \times 1 \times 10^3$.

Solving, we get

$$V_{CE} = 13.96 \text{ volt.}$$
$$\therefore V_E = I_E R_E$$
or $$V_E = (I_C + I_B) R_E = (2 \times 10^{-3} + 40 \times 10^{-6}) \times 1 \times 10^3$$
or $$V_E = 2.04 \text{ volt.}$$

Again $$V_B = V_{CC} - I_B R_B = 20 - 40 \times 10^{-6} \times 430 \times 10^3$$
or $$V_B = 2.8 \text{ volts.}$$
$$\therefore V_C = V_{CE} + V_E = 13.96 + 2.04$$
$$V_C = 16 \text{ volts.}$$

Example 3.22. In the given circuit shown $h_{FE} = 100$, $V_{BE} = 0.8$ volt, $V_{CE} = 0.2$ volt. Determine whether or not the Si transistor is in saturation and find I_B and I_C.

Solution: As $V_{CE} = 0.2$ volt for the Si transistor which is in saturation state, Applying KVL to the input side, we have

$$5 = 5 \times 10^3 I_B + V_{BE} + (I_C + I_B) \times 2 \times 10^3 \quad ...(i)$$

But $$I_C = \beta I_B$$
or $$I_C = h_{FE} \times I_B$$
or $$I_C = 100 I_B \quad ...(ii)$$

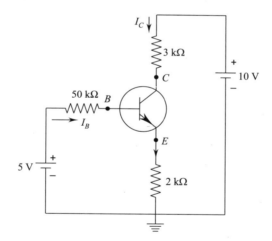

Using equations (*i*) and (*ii*), we get

$$I_B = \frac{5 - 0.8}{5000 + 101 \times 2 \times 10^3}$$

or $$I_B = 16.67 \text{ mA.}$$

Also, $$I_C = 100 I_B$$
$$I_C = 100 \times 16.67 \times 10^{-6}$$
$$= 1.67 \text{ mA.}$$

Now applying KVL to the output side, we get

$$10 = I_C \times 3000 + V_{CE} + (I_C + I_B) \times 2 \times 10^3$$

or $\quad V_{CE} = 10 - 3000 \times 1.67 \times 10^{-3} - (1.67 \times 10^{-3} + 16.67 \times 10^{-6}) \times 2 \times 10^3$

Solving, we get $\quad V_{CE} = 1.617$ volt.

Thus, in the given circuit, we have

$$V_{CE} = 1.617 \text{ volts} > 0.2 \text{ volt.}$$

Therefore, the transistor is not working in saturation region, rather, it is inactive region.

Example 3.23 An Si transistor with $(V_{BE})_{sat} = 0.8$ V, $\beta = h_{FE} = 100$, $(V_{CE})_{sat} = 0.2$ V is used in the circuit shown in the given figure. Find the minimum value of R_C for which the transistor remains in saturation.

Solution: Using KVL in base circuit for saturation, we have

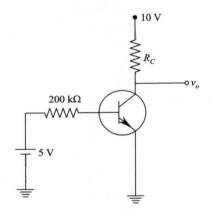

$$V_{BB} = (I_B)_{sat} R_B + (V_{BE})_{sat}$$
$$5 = (I_B)_{sat} \times 200 \times 10^3 + 0.8$$

$$(I_B)_{sat} = \frac{5 - 0.8}{200 \times 10^3}$$

$$= \frac{4.2}{200 \times 10^3} = 21 \times 10^{-6} \text{ amp}$$

Now, we have $\quad (I_C)_{sat} = \beta (I_B)_{sat}$
$$= 100 \times 21 \times 10^{-6} = 2.1 \text{ Amp.}$$

Again, using KVL in collector circuit, we get

$$V_{CC} = (I_C)_{sat} \times R_C + (V_{CE})_{sat}$$
$$10 = 2.1 \times 10^{-3} \times R_C + 0.2$$

Simplifying, we get $\quad R_C = \dfrac{10 - 0.2}{2.1 \times 10^{-3}} = \dfrac{9.8 \times 10^3}{2.1} = 4666$ ohms.

Bipolar Junction Transistor (BJT)

Example 3.24 A transistor with $\beta = 100$ is used in C_E configuration. The collector circuit resistance is $R_C = 1\ k\Omega$ and $V_{CC} = 20$ V. Assuming $V_{BE} = 0$, find the value of collector to base resistance, such that quiescent collector emitter voltage, is 4 V.

Solution: From the figure, it may be observed that

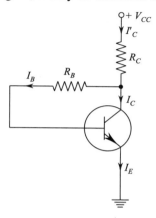

$$I'_C = I_C + I_B$$

∵ $$I_C \gg I_B$$

therefore, $$I'_C = I_C$$

Again from the figure, we have

$$V_{CC} = I_C R_C + V_{CE}$$

Subsituting values, we shall have

$$20 = I_C \times 1 \times 10^3 + 4$$

or $$16 = I_C \times 10^3$$

or $$I_C = 16\ mA$$

We know that $$I_C = \beta I_B$$

or $$I_B = \frac{I_C}{\beta} = \frac{16 \times 10^{-3}}{100} = 160\ mA$$

From the Figure

$$V_{CE} = I_B R_B + V_{BE}$$
$$V_{CE} = I_B R_B + 0$$

$$R_B = \frac{V_{CE}}{I_B} = \frac{4}{160 \times 10^{-6}} = \frac{4}{16 \times 10^{-5}}$$

$$R_B = \frac{1}{4} \times 10^5 = 0.25 \times 10^5$$

$$R_B = 25 \times 10^3$$

$$R_B = 25\ k\Omega.$$

Example 3.25 Find I_C and V_{CE} for the following circuit. What will happen to V_{CE} if β increases due to temperature?

Solution: Applying KVL to the input side, we get

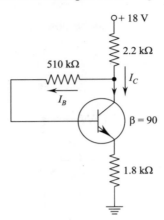

$$V_{CC} = I_C R_C + I_B R_B + V_{BE} + I_E R_E$$

Substituting values, we get

$$18 = 2.23 \times 10^3 I_C + 510 \times 10^3 I_B + 0 + 1.8 \times 10^3 I_E$$

or

$$\frac{18}{1000} = 4 I_E + 510 I_B + 0 \quad ...(i)$$

$\therefore \quad I_E \cong I_C$ and $V_{BE} = 0$ (assuming)

$$\frac{18}{1000} = 4 (I_C + I_B) + 510 I_B$$

But,

$$\beta = \frac{I_C}{I_B} = 90 \text{ or } I_C = 90 I_B \quad ...(ii)$$

Substituting, the value of I_C in equation (i), we have

$$\frac{18}{1000} = 4 \times 90 I_B + 514 I_B$$

$$I_B = \frac{18}{1000 \times 874} = 20.595 \ \mu A$$

Now, value of I_C will be (using equation (ii))

$$I_C = 90 \times 20.95 \text{ m Amp}$$
$$I_C = 1.85 \times 10^{-3} \text{ m amp} = 1.85 \text{ mA.}$$

Again applying KVL to the output side, we get

$$V_{CC} = 18 = 2.2 \times 10^3 I_E + V_{CE} + 1.8 I_E$$
$$18 = 4 \times 10^3 I_E + V_{CE}$$

or $\quad 18 - 4 \times 10^3 (1.85 \times 10^{-3} + 20.595 \times 10^{-6}) = V_{CE}$

or $\quad 18 - 7.4824 = V_{CE}$

or $\quad V_{CE} = 10.52$ volt.

If B increases due to temperature, then V_{CE} will decrease.

Bipolar Junction Transistor (BJT)

Example 3.26 If $I_C = 5$ mA and $I_B = 0.02$ mA, what is current gain ?

Solution: Given $I_C = 5$ mA

and $I_B = 0.02$ mA

Current gain $\beta = \dfrac{I_C}{I_B} = \dfrac{5 \times 10^{-3}}{0.02 \times 10^{-3}}$

$\beta = 250$.

Example 3.27 Draw the load line for the following figure. What is I_C at saturation point? Find V_{CE} at cut-off point.

Solution: We know that

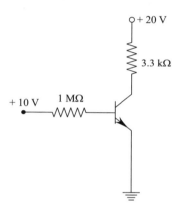

$$V_{CC} = I_C R_C + V_{CE}$$

or $$I_C = -\dfrac{1}{R_C} V_{CE} + \dfrac{V_{CC}}{R_C}$$

It cuts the X-axis on output characteristic at

$V_{CE} = V_{CC}$ { V_{CE} at cut-off}

$V_{CE} = 20$ V

It cuts the Y-axis on output characteristic

$$I_C = \dfrac{V_{CC}}{R_C} = \dfrac{20}{3.3 \text{ k}}$$

$I_C = 6.06$ mA.

Load line can be drawn by joining two points having coordinates $\left(0, \dfrac{V_{CC}}{R_C}\right)$ and $(V_{CC}, 0)$, i.e., (0, 6.06 mA) and (20 V, 0)

Here, we can plot load line as shown in below figure and slope of load line is

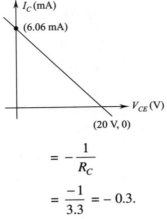

$$= -\frac{1}{R_C}$$

$$= \frac{-1}{3.3} = -0.3.$$

Here the coordinates at Y-axis and X-axis, represent the saturation of I_C and cut-off of V_C respectively.

Example 3.28 Find V_{CE} and I_E in the given figure.

Solution: This is potential divider biasing of transistors, so first let us calculate, the R_{TH} and V_{TH} input circuit, where

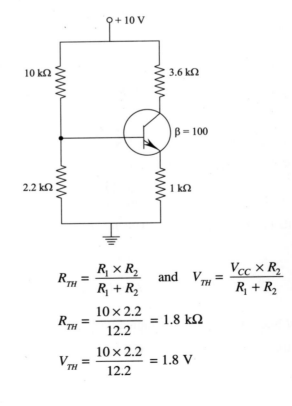

$$R_{TH} = \frac{R_1 \times R_2}{R_1 + R_2} \quad \text{and} \quad V_{TH} = \frac{V_{CC} \times R_2}{R_1 + R_2}$$

Now, $$R_{TH} = \frac{10 \times 2.2}{12.2} = 1.8 \text{ k}\Omega$$

and $$V_{TH} = \frac{10 \times 2.2}{12.2} = 1.8 \text{ V}$$

Applying KVL to input side, we get

$$V_{TH} = I_B R_{TH} + V_{BE} + I_E R_E$$

We know that

$$I_E = (\beta + 1) I_B$$

$$V_{TH} = \frac{I_E}{\beta_H} R_{TH} + V_{BE} + I_E R_E$$

or

$$I_E = \frac{V_{TH} - V_{BE}}{R_E + \frac{R_{TH}}{\beta + 1}}$$

Here, $V_{TH} = 1.8$ V, $V_{BE} = 0.7$ V, $R_E = 1$ kΩ, $R_{TH} = 1.8$ kΩ, $\beta = 100$.
Therefore, we have

So,

$$I_E = \frac{18 - 0.7}{1000 + \frac{1800}{101}} = 1.081 \text{ mA}$$

Now,

$$I_C = \beta I_B = 100 \times 10.7 \text{ mA} = 107 \text{ mA}.$$

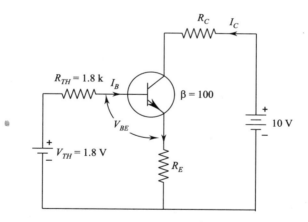

Further, KVL in output loop will yield

$$V_{CC} = I_C R_C - V_{CE} + I_E R_E$$

or

$$V_{CE} = V_{CC} - I_C R_C - I_E R_E$$

or

$$V_{CE} = 10 - 1.07 \times 10^{-3} \times 3.6 \times 10^{-3} - 1.081 \times 10^{-3}$$

$$V_{CE} = 5.067 \text{ volts.}$$

Example 3.29 Determine V_C and I_B for the following network in the given figure.

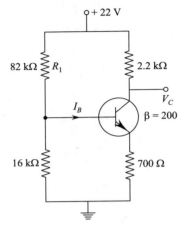

Solution:

Here,
$$V_{TH} = \frac{V_{CC} \times R_2}{R_1 + R_2} = \frac{16 \text{ k}\Omega}{82 \text{ k}\Omega + 16 \text{ k}\Omega} \times (+22)$$

or $V_{TH} = +3.59$ volts.

and
$$R_{TH} = \frac{R_1 \cdot R_2}{R_1 + R_2} = \frac{16 \times 82}{16 + 32} \text{ (k ohm)}$$
$$= 13.39 \text{ k ohm}.$$

Kirchoff's law in input loop will yield

$$V_{TH} = I_E R_E + V_{BE} + I_B R_{TH}$$

We know that $I_E = (\beta + 1) I_B$

and V_{BE} for Si transistor $= 0.7$ V

So, $V_{TH} = (\beta + 1) I_B R_E + V_{BE} + I_B R_{TH}$

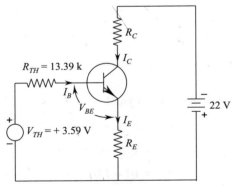

Thevenin's equivalent circuit

or
$$I_B = \frac{V_{TH} - V_{BE}}{R_{TH} + (\beta + 1) R_E}.$$

Here $V_{TH} = +3.59$ V, $R_{TH} = 13.39$ k
$R_E = 750$ ohm
$\beta = 200$
$V_{EB} = 0.7$ volts

So, $I_B = \dfrac{3.59 - 0.7}{13.39 \times 10^3 + (200+1)750}$

or, $I_B = 17.61$ µA (in the shown direction)

We know that,
$$I_C = I_B = 200 \times 17.61 \times 10^{-6} = 3.52 \text{ mA. (in the shown direction)}$$

KVL in the output loop will yield
$V_{CC} = -I_C R_C + V_C$

So, $V_C = V_{CC} + I_C R_C$

Here $V_{CC} = -22$ volts, $R_C = 2.2$ K, $I_C = 3.52$ mA

So, $V_C = -22 + 3.52 \times 10^{-3} \times 2.2 \times 10^3$
$= -14.256$ volts.

Example 3.30 For the emitter follower with $R_S = 0.5$ kΩ and $R_L = 5$ kΩ, calculate A_I, R_i, A_V. Assume $h_{fe} = 50$, $h_{ie} = 1$ kΩ, $h_{oe} = 25$ micro amp/volt.

Solution: For emitter follower,

(i) The current gain, $A_I = \dfrac{1 + h_{fe}}{1 + h_{oe} \cdot R_L} = \dfrac{1 + 50}{1 + 25 \times 10^{-6} \times 5 \times 10^3}$

$A_I = \dfrac{51}{1.125} = 45.33$.

(ii) Input resistance, $R_i = h_{ie} + h_{re} A_I \times R_L$
$R_i = h_{ie} + 1 \times A_I \times R_L$
$R_i = h_{ie} + A_I \times R_L$

Putting all the values, $R_i = 1 \times 10^3 + 45.33 \times 5 \times 10^3$
$R_i = (1 + 226.6) 10^3 = 227.6$ kΩ.

(iii) $A_V = \dfrac{V_o}{V_i} = \dfrac{A_I \cdot R_L}{R_i}$

$= \dfrac{45.33 \times 5}{227.6} = \dfrac{226.6}{227.6} = 0.9958$.

Example 3.31 Find the values of voltage gain, current gain, input resistance and power gain for a common emitter transistor amplifier with $R_L = 1600$ ohm and $R_S = 1$ kΩ. The transistor has $h_{ie} = 1100$ ohm, $h_{fe} = 2.5 \times 10^{-4}$, $h_{oe} = 25$ micro amp/V.

Solution: For common emitter transistor amplifier

(i) Current gain $A_I = -\dfrac{h_{fe}}{1 + h_{oe} \cdot R_L} = -\dfrac{2.5 \times 10^{-4}}{1 + 25 \times 10^{-6} \times 1600}$

$A_I = -\dfrac{2.5 \times 10^{-4}}{1.04} = -2.4 \times 10^{-4}$.

(ii) Input resistance
$$R_i = h_{ie} + h_{re} \times A_I \times R_L$$
$$\cong h_{ie} \text{ (neglecting the factor } h_{re} \times A_I \times R_L)$$
$$R_i \cong 1100 \text{ ohm. Ans.}$$

(iii) Voltage gain
$$A_V = A_I \cdot \frac{R_L}{R_i} = \frac{-2.4 \times 10^{-4} \times 1600}{1100}$$
$$A_V = -\frac{38.4 \times 10^{-4}}{11} = -3.49 \times 10^{-4}.$$

(iv) Power gain
$$= A_I \times A_V$$
$$= (-2.4 \times 10^{-4}) \times (-3.49 \times 10^{-4})$$
$$= 8.37 \times 10^{-8}.$$

Example 3.32 In the following circuit given $\beta_{dc} = h_{FE} = 130$.
(a) Find I_{CQ} and V_{CEQ}
(b) Find A_V and R_{in} for the circuit of part (a) if $h_{fe} = 50$, $h_{ie} = 1$ kW, $h_{re} = 0$, and $h_{oe} = 0$.

Solution:

(a) ∵ Base voltage, $V_B = V_{CC} \times \dfrac{R_2}{R_1 + R_2}$

$$V_B = \frac{8 \times 510}{510 + 510}$$

$$V_B = \frac{18}{2} = 9 \text{ volt}$$

Value of I_C is given by

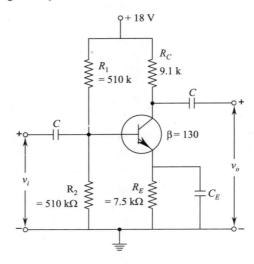

$$I_C = \frac{V_B - V_{BE}}{R_E} = \frac{V_B - 0}{R_E} \text{ (neglecting } V_{BE}\text{)}$$

$$I_C = \frac{V_B}{R_E}$$

$$= \frac{9}{7.5 \times 10^3} = \frac{9 \times 10^{-3}}{7.5}$$

$$I_C = 1.2 \text{ mA}$$

or

$$I_{CQ} = 1.2 \text{ mA}.$$

Now,

$$V_{CE} = V_{CC} - I_C(R_C + R_E)$$
$$= 18 - 1.2 \times 10^{-3}(9.1 + 8.5) \times 10^3$$
$$V_{CE} = 18 - 1.2 \times 16.6 = 18 - 19.9$$
$$V_{CE} = -1.9 \text{ volt or } V_{CEQ} = -1.9 \text{ volt}.$$

(b) Input resistance

$$R_i \cong h_{ie} \cong 1 \text{ k}\Omega.$$

Input resistance of amplifier stage

$$R_{in} = R_i \| (R_1 \| R_2) = \| (510 \| 510)$$
$$R_{in} = \| 255 = 0.99 \text{ k}\Omega.$$

Current gain $\quad A_I \cong h_{fe} = -50$

Voltage gain $\quad A_V = \dfrac{A_I \cdot R_L}{R_i} = \dfrac{A_I \cdot R_C}{R_i} = \dfrac{-50 \times 9.1}{1}$

$$= -455.$$

Example 3.33 Given that $h_{fe} = 50$, $h_{ie} = 0.83$ kΩ. Find out the current gain (h_{fb}) and input impedance (h_{ib}) for a transistor in CB configuration.

Solution: We know that

$$h_{fb} = \frac{-h_{fe}}{1 + h_{fe}}, \text{ substituting the given values,}$$

we have

$$h_{fb} = \frac{-50}{1 + 50} = -0.98.$$

Also,

$$h_{ie} = \frac{h_{ie}}{1 + h_{fe}} = \frac{0.83 \times 10^3}{1 + 50} = 16.27 \ \Omega.$$

Example 3.34 The h-parameters for a CE configuration are $h_{ie} = 2600 \ \Omega$, $h_{fe} = 100$, $h_{re} = 0.02 \times 10^{-2}$ and $h_{oe} = 5 \times 10^{-6}$ S Find h-parameters for CC configuration.

Solution: we know that $h_{ic} = h_{ie}$

It is given that $\quad h_{ie} = 2600 \ \Omega$

Therefore, $\quad h_{ic} = h_{ie} = 2600 \ \Omega.$

Also,
$$h_{fc} = -(1 + h_{fe}) = -(1 + 100)$$
or
$$h_{fc} = -101.$$
and
$$h_{rc} = 1 - h_{rc} \cong 1.$$
$$h_{oc} = h_{oe} = 5 \times 10^{-6} \text{ S}.$$

Example 3.35 A bipolar junction transistor has the following h-parameters: $h_{ie} = 2{,}000\ \Omega$, $h_{re} = 1.6 \times 10^{-4}$, $h_{fe} = 49$; $h_{oe} = 50$ mA/V. Determine the current gain, voltage gain, input resistance and output resistance of the CE amplifier, if the load resistance is 30 kW and the source resistance is 600 Ω.

Solution: We know that

Current gain, $\quad A_i = \dfrac{-h_{fe}}{1 + h_{oe} R_L}$

Substituting the given values, we get

$$A_i = \dfrac{-49}{1 + 50 \times 10^{-6} \times 30 \times 10^3} = -19.6$$

Input resistance, $\quad R_i = h_{ie} - \dfrac{h_{re} \cdot h_{fe}}{h_{oe} + \dfrac{1}{R_L}}$

Substituting the given values, we get

$$R_{in} = 2000 - \dfrac{1.6 \times 10^{-4} \times 49}{50 \times 10^{-6} + \dfrac{1}{30 \times 10^3}} = 1.9062.$$

Also, voltage gain, $\quad A_V = \dfrac{-h_{fe}}{\left[h_{oe} + \dfrac{1}{R_L}\right] R_{in}} = \dfrac{-h_{fe} \cdot R_L}{(1 + h_{oe} R_L) R_{in}} = \dfrac{A_i R_L}{R_{in}}$

Substituting the given values, we get

$$A_V = \dfrac{-19.6 \times 30 \times 10^3}{1.906} = -308.5.$$

Overall voltage gain $\quad A_{VS} = \dfrac{A_V \cdot R_{in}}{R_{in} + R_S}$

or
$$A_{VS} = \dfrac{-308.5 \times 1.906}{1.906 + 600} = -235.$$

Overall current gain $\quad A_{iS} = A_i \dfrac{R_S}{R_{in} + R_S}$

or
$$A_{iS} = \dfrac{-19.6 \times 600}{1.906 + 600} = -4.7.$$

Output conductance, $G_{out} = h_{oe} - \dfrac{h_{fe} \cdot h_{re}}{h_{ie} + R_S}$

or $G_{out} = 50 \times 10^{-6} - \dfrac{49 \times 1.6 \times 10^{-4}}{21000 + 600}$

$G_{out} = 46.985 \times 10^{-6}$ s.

Output resistance, $R_{out} = \dfrac{1}{G_{out}} = \dfrac{1}{46.985 \times 10^{-6}}$

or $R_{out} = 21{,}283 \; \Omega$ or $21.283 \; k\Omega$.

Example 3.36 A BJT has $h_{ie} = 2 \; k\Omega$, $h_{fe} = 100$, $h_{re} = 2.5 \times 10^{-4}$ and $h_{oe} = 25$ mA/V as parameter in CE configuration. It is used as an emitter follower amplifier with $R_s = 1 \; k\Omega$ and $R_L = 500 \; \Omega$. Determine for the amplifier, the voltage gain $AV_s = \dfrac{V_o}{V_s}$, the current gain $A_{is} = \dfrac{I_o}{I_s}$, the input resistance R_i and output resistance R_o.

Solution: For emitter follower (*i.e.*, common-collector amplifier), transistor parameters are given as under:

$$h_{ic} = h_{ie} = 2 \; k\Omega$$
$$h_{fc} = -(1 - h_{fe}) = -(1 + 100) = -101$$
$$h_{rc} = 1 - h_{re} = 1 - 2.5 \times 10^{-4} = 0.99975 \approx 1$$
$$h_{oc} = h_{oe} = 25 \times 10^{-6} \; s$$

Current gain, $A_i = \dfrac{-h_{fc}}{1 + h_{oc} R_L} = \dfrac{-(-101)}{1 + 25 \times 10^{-6} \times 500} = 99.75$

Input resistance, $R_{in} = \dfrac{h_{ic} - h_{rc} \cdot h_{fc}}{h_{oc} + \dfrac{1}{R_L}}$

or $R_{in} = 2 \times 10^3 \; \dfrac{-1 \times (-101)}{25 \times 10^{-6} + \dfrac{1}{500}} = 51.876 \; k\Omega$.

Voltage gain

$A_v = \dfrac{-h_{fc}}{\left[h_{oc} + \dfrac{1}{R_L} \right] R_{in}} = \dfrac{-(-101)}{\left[25 \times 10^{-6} + \dfrac{1}{500} \right] \times 51.876 \times 10^3} = 0.9432$

Overall current gain $A_{is} = A_i - \dfrac{R_s}{R_{in} + R_s}$

or $A_{vs} = 99.75 \times \dfrac{1}{51.876 + 1} = 1.886$

Output conductance $\quad G_o = h_{oc} \dfrac{-h_{fc} \cdot h_{rc}}{h_{ic} + R_s} = 25 \times \dfrac{10^{-6} - (-101) \times 1}{2 \times 10^3 + 1 \times 10^3}$

$G_o = 33.69 \times 10^{-3}$ s

Output resistance, $\quad R_o = \dfrac{1}{G_o} = \dfrac{1}{33.69 \times 10^{-3}} = 29.68\ \Omega.$

Example 3.37 Find A_v and R_{in} for the given circuit as below, if $h_{fe} = 50$, $h_{ie} = 1\ \text{k}\Omega$, $h_{re} = 0$ and $h_{oe} = 0$.

Solution: From figure we have

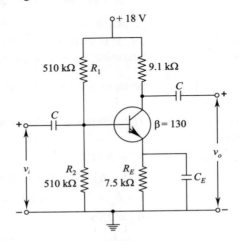

Base voltage $\quad V_B = \dfrac{V_{CC} \times R_2}{R_1 + R_2}$

or $\quad V_B = \dfrac{18 \times 510}{510 + 510}$

$V_B = \dfrac{18}{2} = 9$ volt

Value of current I_C is given by

$I_C = \dfrac{V_B - V_{BE}}{R_E} = \dfrac{V_B - 0}{R_E}$ (neglecting V_{BE})

or $\quad I_C = \dfrac{V_B}{R_E} = \dfrac{9}{7.5 \times 10^3} = \dfrac{9 \times 10^{-3}}{7.5}$

or $\quad I_C = 1.2$ mA

$I_{CQ} = 1.2$ mA

Again from figure, we have

$V_{CE} = V_{CC} - I_C \times (R_C + R_E)$

Bipolar Junction Transistor (BJT)

Substituting values, we get
$$V_{CE} = 18 - 1.2 \times 10^{-3} (9.1 + 7.5) \times 10^3$$
$$= 18 - 1.2 \times 16.6 = -1.9 \text{ volt}$$
$$V_{CEQ} = -1.9 \text{ volt.}$$

Also, input resistance
$$R_i \cong h_{ie} \cong 1 \text{ k}\Omega.$$

Further, input resistance of amplifier stage is given by
$$R_{in} = R_i \parallel (R_1 \parallel R_2) = 1 \parallel (510 \parallel 510) = 1 \parallel 255$$
$$= 0.99 \text{ k}\Omega$$

or current gain is given by
$$A_i = -h_{fe} = -50$$

Voltage gain,
$$A_V = \frac{A_i R_L}{R_i} \text{ but } R_L = R_C, \text{ therefore}$$

$$A_V = \frac{A_i R_C}{R_i} = \frac{-50 \times 9.1}{1} = -455$$

$$A_V = -455.$$

Example 3.38. Find I_{CQ} and I_{CEQ} for the given circuit shown below, given that $\beta_{dc} = h_{FE} = 130$.

Solution: From figure, we have
$$R_{TH} = \frac{R_1 \times R_2}{R_1 + R_2} = \frac{510 \times 510}{510 + 510} \quad [\because R_1 \text{ and } R_2 \text{ in } \parallel]$$

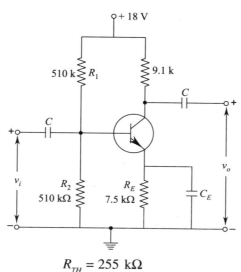

or
$$R_{TH} = 255 \text{ k}\Omega$$

and
$$V_{TH} = \frac{V_{CC} \times R_2}{R_1 + R_2} = \frac{510 \times 18}{510 + 510} = 9 \text{ volt}$$

Therefore, we have $I_B = \dfrac{9-0.7}{255+(130+1)\times 7.5} = \dfrac{8.3}{255+131\times 7.5}$

or $\quad I_B = \dfrac{8.3}{1237.5}$

or $\quad I_B = 0.00670$ mA

We know that $\quad I_C = \beta I_B = 130 \times 6.7 \times 10^{-3}$

$I_C = 0.8719$ mA.

Again, from figure, we get

$$V_{CE} = V_{CC} - I_C \times R_C - I_E R_E$$
$$V_{CE} = 18 - 0.8719 \times 91 - (0.8719 + 0.0067)\times 7.5$$
$$V_{CE} = 3.476 \text{ volt.}$$

Example 3.39 A transistor has α_{dc} of 0.98 and a collector leakage current I_{CO} of 1 mA.

Calculate the collector and the base current when $I_E = 1$ mA

Solution: Width $\quad I_E = 1$ mA

We can use equation $\quad I_C = \alpha_{dc} I_E + I_{CO}$

$= 0.98 \times 1 \times 10^{-3} + 1 \times 10^{-6} = 0.981 \times 10^{-3}$

$= 0.981$ mA

Now using equation $\quad I_B = I_E - I_C$

$= 1\times 10^{-3} - 0.981 \times 10^{-3}$

$= 0.019 \times 10^{-3} = 0.019$ mA

$= 19$ mA

Note that I_C and I_E are almost equal and I_B is very small.

Example 3.40 In a transistor, a change in emitter current of 1 mA produces a change in collector current of 0.99 mA. Determine the short circuit current gain of the transistor.

Solution: The short circuit current gain of the transistor is given as

$$\alpha \quad \text{or} \quad h_{fb} = \dfrac{\Delta_{iC}}{\Delta_{iE}} = \dfrac{0.99 \times 10^{-3}}{1 \times 10^{-3}} = 0.99.$$

Example 3.41 When the emitter current of transistor is changed by 1 mA, its collector current changes by 0.995 mA. Calculate (a) its common-base short circuit current gain α and (b) its common-emitter short circuit current gain β.

Solution:

(a) Common-base short circuit current gain is given by

$$\alpha = \dfrac{\Delta_{iC}}{\Delta_{iE}} = \dfrac{0.995 \times 10^{-3}}{1 \times 10^{-3}} = 0.995$$

(b) Common-Emitter short circuit current gain is

$$\beta = \frac{\alpha}{1-\alpha} = \frac{0.995}{1-0.995} = 199.$$

Example 3.42 The dc current gain of a transistor in common emitter configuration is 100. Find its dc current gain in common-base configuration.

Solution: We can use the given equation to calculate the dc current gain in common base configuration.

$$\alpha_{dc} = \frac{\beta_{dc}}{\beta_{dc}+1} = \frac{100}{100+1} = 0.99$$

Example 3.43 Calculate the collector current and the collector-to-emitter voltage for the circuit given as follows:.

Solution:

(a) The base current I_B is given as

$$I_B = \frac{(V_{CC}-V_{BE})}{R_B} \approx \frac{V_{CC}}{R_B} = \frac{9}{300 \times 10^{-3}}$$

$$= 3 \times 10^{-5} \text{ A} = 30 \text{ μA}.$$

(b) The collector current I_C is given

$$I_C = \beta I_B = 50 \times 30 \times 10^{-6} \text{ A}$$

$$= 1.5 \text{ mA}.$$

Let us check if this current is less than the collector saturation current

$$I_{C \text{ (sat)}} = \frac{V_{CC}}{R_C} = \frac{9}{2 \times 10^3} = 4.5 \times 10^{-3} \text{ A} = 4.5 \text{ mA}$$

Thus, the transistor is not saturated

(c) The collector-to-emitter voltage

$$V_{CE} = V_{CC} - I_C R_C = 9 - 1.5 \times 10^{-3} \times 2 \times 10^3 = 6 \text{ V}$$

Example 3.44 Calculate the coordinates of the operating point as fixed in the given circuit shown below. Given $R_C = 1$ kΩ, $R_B = 100$ kΩ.

Solution:

(a) The base current is

$$I_B = \frac{V_{CC}-V_{BE}}{R_B} \approx \frac{V_{CC}}{R_B} = \frac{10}{100 \times 10^3} \text{ A} = 100 \text{ μA}$$

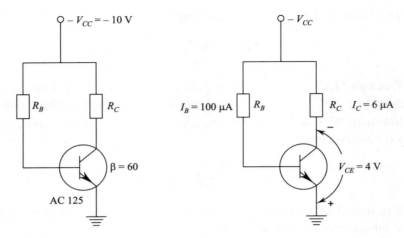

(*b*) The collector current is

$$I_C = \beta I_B = 60 \times 100 \times 10^{-6} \text{ A} = 6 \text{ mA}$$

We shall not check if this current is less than the collector saturation current

$$I_{C\text{ (sat)}} = \frac{V_{CC}}{R_C} = \frac{10}{1 \times 10^3} \text{ A} = 10 \text{ mA}.$$

Therefore, the transistor is not in saturation.

(*c*) The voltage between the collector and emitter terminals is

$$V_{CE} = V_{CC} - I_C R_C = 10 - 6 \times 10^{-3} \times 10^3 = 4 \text{ V}$$

Figure shows the value and the direction of base current I_B, collector current I_C and collector-emitter voltage V_{CE}.

Example 3.45 In the given circuit following figure, the transistor is replaced by another unit of AC 125. This new transistor has $\beta = 150$ instead of 60. Determine the quiescient operating point.

Solution:

(*a*) The base current remains the same, *i.e.*, 100 mA.

(*b*) The collector current is

$$I_C = \beta I_B = 150 \times 100 \times 10^{-6} \text{ A} = 15 \text{ mA}$$

The collector saturation current was 10 mA in the last example. Here also, this current remains the same. But the calculated current I_C is seen to be greater than $I_{C\text{ (sat)}}$. Hence, the transistor is now in saturation. In this case, the operating point is specified as

$$I_C = I_{C\text{ (sat)}} = 10 \text{ mA}$$
$$V_{CE} = 0 \text{ V}.$$

Bipolar Junction Transistor (BJT)

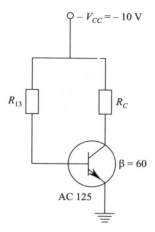

Example 3.46 How much is the emitter current in the following circuit, also calculate V_C.

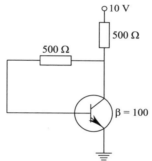

Solution: From given equation, the base current is given as

$$I_B = \frac{V_{CC}}{R_B + \beta R_C}$$

Here, $V_{CC} = 10$ V ; $R_B = 500 \times 10^3$ Ω
$R_C = 500$ Ω, $\beta = 100$

Therefore,

$$I_B = \frac{10}{500 \times 10^3 + 100 \times 500}$$

$$= 18 \times 10^{-6} \text{ A} \neq 18 \text{ mA}$$

The emitter current is then given as

$$I_C \cong I_C = \beta I_B = 100 \times 18 \times 10^{-6} = 1.8 \times 10^{-3} \text{ A} = 1.8 \text{ mA}$$

The collector vol

$$V_C = V_{CE} = V_{CC} - I_C R_C = 10 - 1.8 \times 10^{-3} \times 500 = 9.1 \text{ V}.$$

Example 3.47 Calculate the value of the 3 currents in the following circuit:
Solution: From the equation given, the base current is given as

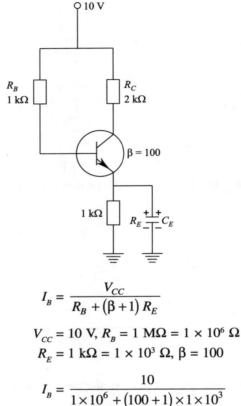

$$I_B = \frac{V_{CC}}{R_B + (\beta + 1) R_E}$$

Here
$V_{CC} = 10$ V, $R_B = 1$ MΩ $= 1 \times 10^6$ Ω
$R_E = 1$ kΩ $= 1 \times 10^3$ Ω, $\beta = 100$

Therefore,
$$I_B = \frac{10}{1 \times 10^6 + (100 + 1) \times 1 \times 10^3}$$

$= 9.09 \times 10^{-6}$ A

$= 9.09$ mA

Now, the collector current
$$I_C = \beta I_B = 100 \times 9.09 \times 10^{-6}$$
$= 0.909 \times 10^{-3}$ A
$= 0.909$ μA

The emitter current
$$I_E = I_C + I_B \approx I_C = 0.909 \text{ mA}.$$

Example 3.48 If the collector resistance R_C in given cicuit is changed to 1 kW, determine the new Q points for the minimum and maximum values of β.

Solution: Since the value of the emitter current does not depend upon the value of R_C equation (*i*), the emitter current I_E remains the same as calculated in previous example. That is,

Bipolar Junction Transistor (BJT)

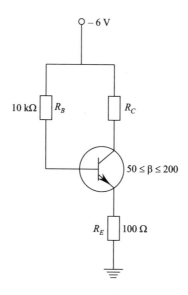

$$I_E = \frac{(V_{CC} - V_{BE})(\beta + 1)}{R_B + (\beta + 1) R_E}$$

(i) For $\beta = 50$, $I_E = 19.25$ mA
(ii) For $\beta = 200$, $I_E = 38.2$ mA.

In case (i), the collector to emitter vol. is given by

$$V_{CE} = V_{CC} - (R_C + R_E) I_E = 6 - (1000 + 100) \times 19.25 \times 10^{-3}$$
$$= 6 - 21.17 = -15.17 \text{ V}.$$

The above result is absurd ! sum of the V drop across R_C and R_E cannot be greater than the supply vol. V_{CC}. Is our calculation wrong ? Certainly not, we face such difficulties when the transistor is in saturation. The maximum possible current that can be supplied by the battery V_{CC} to the output section is

$$I_{C\text{ (sat)}} = \frac{V_{CC}}{R_C + R_E} = \frac{6}{1000 + 100} = 5.45 \times 10^{-3} \text{ A} = 5.45 \text{ mA}$$

Under saturation, the collector-to-emitter vol. is

$$V_{CE\text{ (sat)}} = 0 \text{ V}$$

(ii) We have seen that the transistor is in saturation when its $\beta = 50$. In case $\beta = 200$, there is all the more reason for the transistor to be in saturation. So, the Q point will be the same as calculated earlier, i.e.,

$$I_{C\text{ (sat)}} = 5.45 \text{ mA}, V_{C\text{ (sat)}} = 0 \text{ V}$$

Example 3.49 Calculate the value of R_B in the biasing circuit of given below circuit so that the Q point is fixed at $I_C = 8$ mA and $V_{CE} = 3$ V.

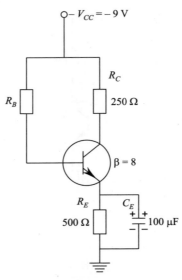

Solution: The current I_B is given as

$$I_B = \frac{I_C}{\beta}$$

Here, $I_C = 8$ mA $= 8 \times 10^{-3}$ A and $\beta = 80$

Therefore, $$I_B = \frac{8 \times 10^{-3}}{80} = 1 \times 10^{-4} \text{ A} = 100 \text{ mA}$$

From equation, $$I_B = \frac{V_{CC} - V_{BE}}{R_B + (\beta + 1) R_E} \simeq \frac{V_{CC}}{R_B + \beta R_E}$$

We have

$$I_B R_B + (\beta + 1) I_B R_E = V_{CC} - V_{BE} \simeq V_{CC}$$

or $$R_B = \frac{V_{CC} - (\beta + 1) I_B R_E}{I_B}$$

Here $V_{CC} = 9$ V, $\beta = 80$, $I_B = 1 \times 10^{-4}$ A, $R_E = 500$ Ω

Therefore, $$R_B = \frac{9 - (80 + 1) \times 1 \times 10^{-4} \times 500}{1 \times 10^{-4}} = \frac{4.95}{1 \times 10^{-4}} \text{ Ω}$$

$$= 49.5 \text{ kΩ}.$$

Bipolar Junction Transistor (BJT)

Example 3.50 To set up 100 mA of emitter current in the power amplifier circuit of given figure.

Calculate the value of the resistor R_E. Also calculate V_{CE}. The dc resistance of the primary of the output transformer is 20 Ω.

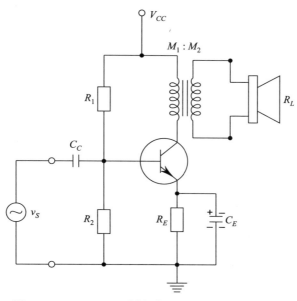

Solution: Given $\quad R_1 = 200\ \Omega,\ R_2 = 100\ \Omega,\ R_C = 20\ \Omega,\ V_{CC} = 15\ V$;
$$I_C \approx I_E = 100\text{ mA} = 0.1\text{ A}$$

From equivalent the base voltage is
$$V_B = \frac{V_{CC} \times R_2}{R_1 + R_2}$$

$$= \frac{100}{200 + 100} \times 15 = 5\text{ V}$$

Neglecting V_{BE}, $\quad V_E = 5\text{ V}$

From equivalent $\quad I_E = \dfrac{V_E}{I_E}$

$$= \frac{5}{0.1} = 50\ \Omega$$

The collector-to-emitter voltage is then calculated using equation
$$V_{CE} = V_{CC} - (R_C + R_E)\,I_C$$
$$= 15 - (20 + 50) \times 0.1 = 8\text{ V}.$$

Example 3.51 Calculate I_C and V_{CE} the emitter-bias circuit of given figure, where $V_{CC} = 12$ V, $V_{BE} = 15$ V, $R_C = 5$ kΩ, $R_E = 10$ kΩ, $R_B = 10$ kΩ, β = 100.

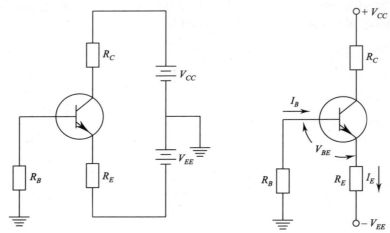

Solution: From equation, $I_E = \dfrac{V_{EE}}{R_E}$, the emitter current is

$$I_E = \dfrac{V_{EE}}{R_E} = \dfrac{15}{10 \times 10^3} = 1.5 \times 10^{-3} \text{ A} = 1.5 \text{ mA}$$

The collector current is $I_C \cong I_E = 1.5$ mA

Using equation $V_{CE} = V_{CC} - I_C R_C$, the voltage V_{CE} is

$$= 12 - 1.5 \times 10^{-3} \times 5 \times 10^3$$
$$= 12 - 7.5 = 4.5 \text{ V}.$$

SUMMARY

1. **Basics:** Bipolar Junction Transistor (BJT) is just called a transistor. It is a three terminal device namely emitter, base and collector. It has three semiconductor layers having a base or centre layer, a great deal thinner than the other two layers. The outer two layers are both of either N or P type materials, and the sandwitched layer the opposite type. The arrow in the transistor symbol defines the direction of conventional current flow for the emitter current and thereby defines the direction for the other currents of the device. The arrow in the symbol of an NPN transistor points out of the device, whereas the arrow points into the centre of symbol for PNP tansistor.

2. **Biasing:** In the active region of a transistor, the base-emitter junction is forward-biased and the collector-base junction is reverse-biased. In the cut-off region the base-emitter and collector-base junctions of a transistor are both reverse-biased. In the saturation region the base emitter and collector-base junctions are forward biased.

Bipolar Junction Transistor (BJT)

3. **Currents:** The dc emitter current is always the largest current of a transistor and the base current is always the smallest. The emitter current is always the sum of the other two. The collector current is made up of two components—the majority component and the minority current also called the leakage current.
4. **V_{BE}, α and β:** Base-to-emitter voltage (V_{BE}) is 0.7 V approximately. α is always close to one. β is usually between 50 and 400.
5. **Impedance between transistor terminals:** The impedance between terminals of forward-biased junction is always relatively small, whereas the impedance between terminals of reverse-biased junction is usually quite large.
6. **Important Formulae:**

   (i) $I_E = I_C + I_B$

   (ii) $V_{BE} \approx 0.7$ V for silicon transistor

   ≈ 0.3 V for germanium transistor

   (iii) $\alpha = \alpha_{dc} = \dfrac{I_C}{I_E}$

   (iv) $\alpha_{ac} = \dfrac{\Delta I_C}{\Delta I_E}\bigg|_{V_{CB} = \text{constant}}$

   (v) $I_{CEO} = \dfrac{I_{CBO}}{1-\alpha}\bigg|_{I_B = 0\mu A}$

   (vi) $\beta = \beta_{dc} = \dfrac{I_C}{I_B}$

   (vii) $\beta_{ac} = \dfrac{\Delta I_C}{\Delta I_B}\bigg|_{V_{CE} = \text{constant}}$

   (viii) $\alpha = \dfrac{\beta}{\beta + 1}$

   (ix) $I_C = \beta I_B + I_{CEO} \approx \beta I_B$ for $I_{CEO} \ll I_B$

   (x) $I_E = (\beta + 1) I_B$

   (xi) $P_{C\max} = V_{CE} I_C$

   (xii) $r_i = \dfrac{\Delta V_{EB}}{\Delta I_E}\bigg|_{V_{CB} = \text{constant}}$ for CB configuration

   (xiii) $r_o = \dfrac{\Delta V_{CB}}{\Delta I_C}\bigg|_{I_E = \text{constant}}$ for CB Configuration

   (xiv) $r_i = \dfrac{\Delta V_{BE}}{\Delta I_B}\bigg|_{V_{CE} = \text{constant}}$ for CE Configuration

   (xv) $r_o' = \dfrac{\Delta V_{CE}}{\Delta I_C}\bigg|_{I_B = \text{constant}}$

(xvi) $R_B = \dfrac{V_{CC} - V_{BE}}{I_B}$ for base resistor method.

(xvii) $R_B = \dfrac{V_{CC} - V_{BE} - \beta I_B I_C}{I_B}$ for feedback resistor

(xviii) $I_C = \dfrac{V_2 - V_{BE}}{R_E}$,

where $V_2 = \left(\dfrac{V_{CC}}{R_1 + R_2}\right) R_2$

and $V_{CE} = V_{CC} - I_C (R_C + R_B)$ for potential divider

(xix) $A_v = \beta \times \dfrac{R_{AC}}{R_i}$, where R_{AC} = ac load

$= R_C$ if no load connected

$= R_C \parallel R_L$ if load R_L connected.

(xx) $A_p = \beta^2 \times \dfrac{R_{AC}}{R_i}$ = Current gain × Voltage gain

(xxi) AC load line is between $V_{CE\,max}$ to $I_{C\,max}$

$V_{CE\,max} = V_{CE} + I_C R_{AC}$

$I_{C\,max} = I_C + \dfrac{V_{CE}}{R_{AC}}$

EXERCISES

3.1 Why is an ordinary transistor called BJT ? Explain basic construction and operation of a transistor.

3.2 Give brief operations of input/output characteristics of a transistor.

3.3 Derive the relationship between α and β.

3.4 What are the factors responsible for the stability of operating point ?

3.5 Draw a self bias circuit and derive an expression for stability factor.

3.6 Static various methods of improving stability.

3.7 Define stability factor w.r.t. transistor biasing.

3.8 Explain the function of emitter in the operation of a junction transistor.

3.9 Derive the hybrid or *h* parameters for a two port network.

3.10 Find the h_{ic} in terms of the CB *h*-parameters.

3.11 In a common base connection, current amplification factor is 0.9. If the emitter current is 1 mA, determine the base current. [**Ans.** 0.1 mA]

3.12 In a common base connection, the emitter current is 1 mA. If the emitter circuit is open, the collector current is 50 μA. Find the total collector current given that α = 0.92. [**Ans.** 0.97 mA]

3.13 Find the value of β if

(i) α = 0.9 (ii) α = 0.98 (iii) α = 0.99 [**Ans.** 9,49,99]

Bipolar Junction Transistor (BJT)

3.14 The base current in a transistor is 0.01 mA and emitter current is 1 mA. Calculate the various of α and β. **[Ans. 0.99, 99]**

3.15 The collector leakage current in a transistor is 300 μA in CE arrangement. If the transistor is now connected in CB arrangement, what will be the leakage current? Given that $\beta = 120$. **[Ans. 2.4 μA]**

3.16 For a certain transistor, $I_B = 20$ μA, $I_C = 2$ mA and $\beta = 80$. Calculate I_{CBO}. **[Ans. 0.0008 mA]**

3.17 In the following circuit, find the operating point given that $\beta = 100$. Neglect V_{BE}.

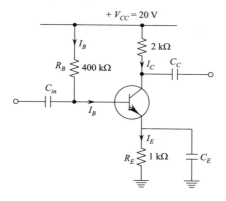

[Ans. 8 V, 4 mA]

3.18 In the given figure shown a silicon transistor biased by feedback resistor method. Determine the operating point, Given that $\beta = 100$.

[Ans. 10.4 V, 9.6 mA]

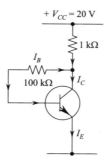

3.19 Find the operating point in the circuit shown below. Assume $\beta = 75$ and $V_{BE} = 0.7$ V. **[Ans. 10.59 V, 2.47 mA]**

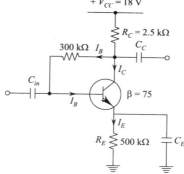

3.20 Given figure shows the voltage divide bias method. Draw the dc load line and determine the operating point. Assume the transistor to be of silicon.

[**Ans.** 8.55 V, 2.15 mA]

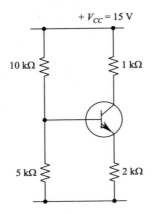

3.21 An NPN transistor circuit has $\alpha = 0.985$ and $V_{BE} = 0.3$ V. If $V_{CC} = 16$ V, calculate R_1 and R_C to place Q point at $I_C = 2$ mA, $V_{CE} = 6$ V.

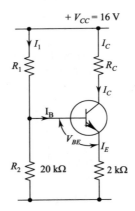

3.22 What are *CB*, *CE* and *CC* configurations of a transistor ? Explain with circuit diagram.

3.23 What do you understand by biasing of a transistor ? Explain various methods for biasing a transistor.

3.24 Give comparison of various methods of biasing a transistor.

3.25 What is graphical analysis of a transistor ? Describe in detail.

3.26 Give the concept of voltage gain and current gain of a transistor.

❑❑❑

4

FIELD EFFECT TRANSISTOR (FET)

4.1 INTRODUCTION

In the bipolar junction transistor, the output collector current dependent on the amount of current flowing into the base terminal, therefore, it is known as a current operated device. Field Effect Transistors or FETs use the voltage which is applied to the gate and source terminals to control the output current. Hence, FET is a voltage operated device. The operation relies on the electric field generated by the input voltage, hence, the name is field effect transistor (FET) and it also implies that FET is voltage operated device. As it is voltage controlled device similar to a vacuum tube, hence, other replaced vaccum tubes.

FETs are unipolar devices which have very similar properties as that of BJTs. FETs also have high efficiency and instant operation, there are also robust, cheap and can be used in most of the applications where BJTs are used. FETs can be made much smaller than an equivalent BJT. It also has lower power consumption and dissipation, hence, they are ideal for use in integrated circuits and computer circuit chips. As an amplifier, the JFET offers a higher input impedance than JBT, generates less self-noise and has greater resistance to nuclear radiations. Thus, **JFETs** have also replaced vacuum tubes.

FETs have extremely high input impedance, hence, these are very sensitive. This also implies that FETs can be damaged by static electricity. FETs are of two major types, the Junction Field Effect Transistor (JFET) and the Metal Oxide Semiconductor Field Effect Transistor (MOSFET) which are also called the Insulated-Gate Field-Effect Transistor (IGFET).

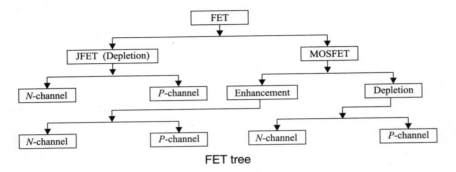

FET tree

4.2 JUNCTION FIELD EFFECT TRANSISTOR (JFET)

A BJT is constructed using *P-N* junctions on the current path between Emitter and Collector terminals. The FET has no junction instead has a narrow "channel" of *N*-type or *P*-type silicon with electrical connections at either and commonly called the Drain and the source. Fig. 4.1 shows the basic construction of the *N*-channel JFET. The major part of the structure is then-type material which forms the channel between the embedded layers of *P*-type material. The top of the *N*-type channel is enacted through an ohmic intact to a terminal known as the drain (D). The lower end of the same material is enacted through an ohmic intact to a terminal known as sure (5). Both *P*-type materials are connected together and both through the gate (G). In short, the drain and the source are connected to the ends of the *N*-type channel and the gate to the two layers of *P*-type materials. If no potential is applied, no-bias renditions. This loads to a depletion region at each junction which looks similar to the same region of a diode under no-bias renditions. Similarly, *P*-channel JFET is as per Fig. 4.2, please bite-that in this case the major part of construction is the *P*-type material. Further, the two *P-N* junctions. in terming diodes are connected internally and gate. The important analogy between BJT and JFET are:

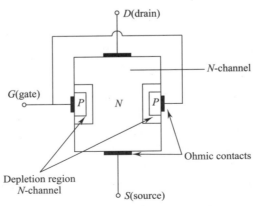

Fig. 4.1 Construction of *N*-channel JFET.

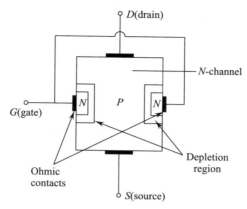

Fig. 4.2 Construction of P-channel JFET.

Bipolar Junction Transistor (BJT)	Junction Field Effect Transistor (JFET)
Emitter (E)	Source (S)
Base (B)	Gate (G)
Collector (C)	Drain (D)

Instead of an AC current a JFET has a dc source current I_S instead of a dc base current I_B, it has a dc drain current I_D.

In a JFET, there is only one type of carrier, holes in P-type channel and electrons in N-type channel. Hence, FET is also known as unipolar. In the ordinary transistor, both holes and electrons play part in construction. Therefore, ordinary transistors are also known as bipolar transistor.

4.3 WORKING PRINCIPLE OF JFET

Polarities of N-channel JFET and P-channel JFET are shown in Fig. 4.3.

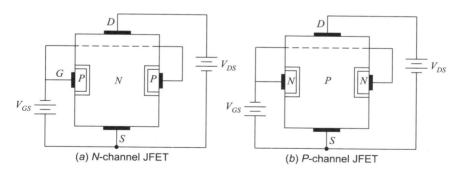

Fig. 4.3 Biased JFET

The voltage between the gate and source is as such that the gate is reverse biased. The drain and source terminal are interchangeable. D and S terminals are interchangeable. That is why, polarity of bias voltage between D and S has not changed in both types of JFET.

Suppose a V_{DS} voltage is applied between drain and source terminals while there is no voltage on the gate as shown in Fig. 4.4(a), the two P-N junctions at the sides of the bar establish depletion layers. Hence, electrons will flow from source to drain through a channel between the depletion layers, the size of the depletion layers determine the width of the channel, therefore, the amount of electron flow through the channel. If a reverse voltage V_{GS} is applied between gate and source as shown in Fig. 4.4(b), the width of the depletion layer is increased. In turn, the width of conducting channel is reduced and as such the resistance of N-type bar is increased. This leads to reduction in current from source to drain. However, if the reverse voltage on the gate is decreased, the width of depletion is alo reduced, thereby increasing the width of the conducting channel. This in turn decreases, the resistance of N-type bar which increases the current from source to drain.

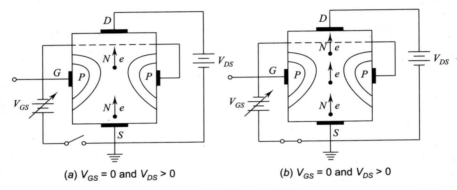

(a) $V_{GS} = 0$ and $V_{DS} > 0$ (b) $V_{GS} = 0$ and $V_{DS} > 0$

Fig. 4.4 Working of JFET.

In short, the current from source to drain can be controlled by applying voltage on the gate. A P-channel JFET operates similar to an N-channel JFET with only one exception that the current carrier will be holes instead of electrons, and also the polarities of V_{GS} and V_{DS} are reversed.

JFET symbol are shown in Fig. 4.5.

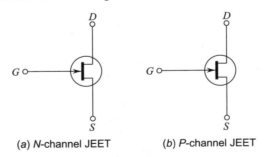

(a) N-channel JEET   (b) P-channel JEET

Fig. 4.5 JFET Symbols.

The arrow points in the direction I_G current would flow of the P-N junction is forward biased. The arrow points in for N-channel JFET whereas arrow points out for P-channel.

4.4 CONCEPT OF PINCH-OFF AND MAXIMUM DRAIN SATURATION CURRENT

Consider JFET of Fig. 4.6, when V_{DS} voltage is increased from zero to a few volts, the current will increase linearly as per Ohm's law. The plot of I_D versus V_{DS} for $V_{GS} = 0$ as shown in Fig. 4.7. When V_{DS} approaches a level V_P of

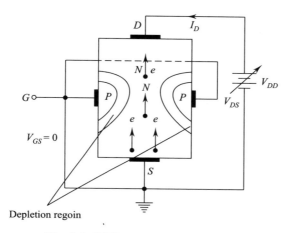

Fig. 4.6 JFET at $V_{GS} = 0$ and $V_{DS} > 0$.

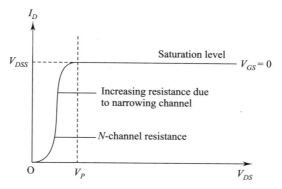

Fig. 4.7 I_D versus V_{DS} plot for $V_{GS} = 0$ V

In Fig. 4.7, the depletion regions widen causing reduction in channel width. This increases the resistance of the current, and finally-there is a situation that when V_{DS} goes beyond a value of V_P, the width of conduction does not reduce further and current density is very high. Further increase of V_{DS} does not affect the channel width and the I_D remains at saturation level. The voltage V_P is known as **pinch-off** voltage. I_{DSS} is the maximum drain current for a JFET is defined by the conditions $V_{GS} = 0$ and $V_{DS} > |V_P|$.

4.5 INPUT AND TRANSFER CHARACTERISTICS

The relationship between input and output of JFET is not linear. The relationship between V_{GS} (input) and I_D (output) is defined by Shockley's characteristics equation given under:

$$I_D = I_{DSS}\left(1 - \frac{V_{GS}}{V_P}\right)^2 \qquad ...(4.1)$$

where, I_{DSS} = a constant value
          V_P = a constant value.

The squared term in the equation gives a non-linear relationship between V_{GS} and I_D. This produces a curve. I_D increases exponentially with decreasing magnitude of V_{GS}. The transfer characteristics given by the equation are unaffected by the network in which the device is employed. The input V_{GS} versus characteristics as shown in Fig. 4.8. I_D is maximum (I_{DSS}) when $V_{GS} = 0$. I_D is zero, when $V_{GS} = 4$ V. The transfer characteristics are obtained for various values of V_{GS} as shown in the right hand side of the plot in Fig. 4.8.

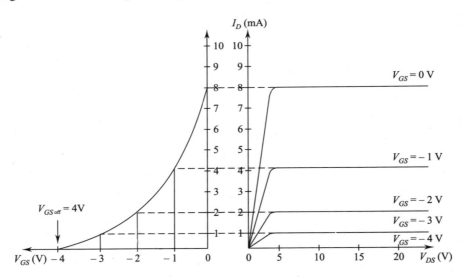

Fig. 4.8 Input and transfer characteristics.

The important aspects of the characteristics may be noted i.e., initially, the drain current I_D increases rapidly with drain-source voltage V_{DS}, but finally it becomes constants.

The gate-source voltage where the channel is completely cut-off and drain current becomes zero is known as Gate Source Cut-off Voltage ($V_{GS\ OFF}$).

4.6 PARAMETERS OF JFET

The main parameters of JFET are ac drain resistance, transconductance and amplification factor.

(i) AC Drain Resistance (r_d)

The ratio of change in drain-source voltage (ΔV_{DS}) to the corresponding change in drain current (ΔI_D) at constant gate-source voltage, is known as ac drain resistance (r_d)

ac drain resistance, $r_d = \dfrac{\Delta V_{DS}}{\Delta I_D}$ at a constant V_{GS}.

It is expressed in kΩ or MΩ. ...(4.2)

(ii) Transconductance (g_m)

The ratio of change in drain current (ΔI_D) to the corresponding change in gate-source voltage (V_{GS}) at a constant drain-source voltage V_{DS}, is known as transconductance (g_m).

Tranconductanec, $g_m = \dfrac{\Delta I_D}{\Delta V_{GS}}$ at a constant V_{DS} ...(4.3)

It is expressed in ma/V, millimhos or micromhos.

(iii) Amplification Factor (μ)

The ratio of change in drain-source voltage (ΔV_{DS}) to the change in gate-source voltage (ΔV_{GS}) at a constant drain current, is known as amplification factor (μ).

Amplification factor, $\mu = \dfrac{\Delta V_{DS}}{\Delta V_{GS}}$ at a constant I_D ...(4.4)

If basically, it gives gate voltage effectiveness compared to the drain voltage in controlling the drain current. It can also be expressed in terms of ac drain resistance and transconductance.

It is known that:

$$\mu = \dfrac{\Delta V_{DS}}{\Delta V_{GS}} \qquad I_D = I_{DSS}\left(1 - \dfrac{V_{GS}}{V_P}\right)^2$$

The above expression can be manipulated as:

$$\frac{\partial I_D}{\partial V_{GS}} = 2 I_{DSS}\left(1-\frac{V_{GS}}{V_P}\right)\left(-\frac{1}{V_P}\right)$$

$$\mu = \frac{\Delta V_{DS}}{\Delta V_{GS}} \times \frac{\Delta I_D}{\Delta I_D} = \frac{2}{|V_P|} I_{DSS}\left(\frac{I_D}{I_{DSS}}\right)^{1/2}$$

$$= \frac{\Delta V_{DS}}{\Delta I_D} \times \frac{\Delta I_D}{\Delta V_{GS}} = \frac{2}{|V_P|}(I_D I_{DSS})^{1/2}$$

$$\therefore \quad \mu = r_d \times g_m \text{ as } \frac{\Delta V_{DS}}{\Delta I_D} = r_d \text{ and } \frac{\Delta F_D}{\Delta V_{GS}} = g_m.$$

$$g_m = g_{m_0}\left(1-\frac{V_{GS}}{V_P}\right)$$

$$\text{where } g_{m_0} = \frac{2 I_{DSS}}{V_P}.$$

4.7 JFET BIASING

JFET gate must be negative with respect to source for the proper operation. A battery in gate circuit or a biasing circuit is essential. **JFET biasing** circuit is preferred as batteries are costly and require frequent replacement.

4.7.1 Fixed-biasing of JFET

Proper gate-source voltage V_{GS} is required to give desired drain current I_D. A fixed biasing is achieved through batteries as shown in Fig. 4.9.

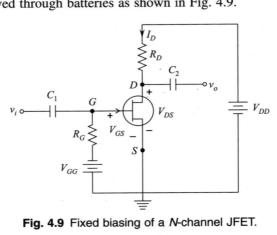

Fig. 4.9 Fixed biasing of a N-channel JFET.

For dc analysis capacitors work as open circuit ($X_c = \dfrac{1}{2\pi f C} = \infty$ as $f = 0$)

From the gate-source voltage circuit, we get:

$$V_{GS} = -V_G - V_S = -V_{GG} - 0$$

or
$$V_{GS} = -V_{GG} \qquad \ldots(4.6)$$

The drain current gets fixed by equation (4.1) i.e.,

$$I_D = I_{DSS}\left(1 - \dfrac{V_{GS}}{V_P}\right)^2.$$

The voltage drop drain resistor R_D is:

$$V_{RD} = I_D R_D$$

∴ Output voltage, $\quad V_o = V_{DD} - I_D R_D \qquad \ldots(4.7)$

4.7.2 Self-biasing of JFET

A **Self-biasing** circuit is shown in Fig. 4.10, R_S is the bias register.

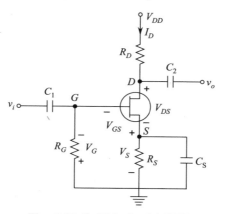

Fig. 4.10 Self-biasing of JFET.

The drain current dc component flowing through resistor R_S creates the desired biasing voltage. The ac component of drain current gets by-passed through C_S capacitor voltage across source resistor R_S is given by:

$$V_S = I_D R_S$$

∴ $\quad V_{GS} = V_G - V_S = 0 - V_S$ as gate current is negligible $V_G = 0$

or $\quad V_{GS} = -I_D R_S$

The above equation keeps the gate negative w.r.t. source to terminal.

The dc operating i.e., zero signal I_D and V_{DS} can be determined by following equations:

$$I_D = I_{DSS}\left(1 - \dfrac{V_{GS}}{V_P}\right)^2$$

and $\quad V_{DS} = V_{DD} - I_D(R_D + R_S)$

4.7.3 Potential-divider Method of Biasing JFET

Potential-divider method of biasing JFET is shown in the circuit of Fig. 4.11.

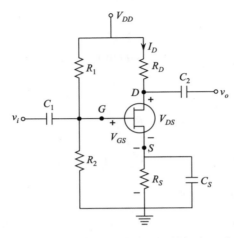

Fig. 4.11 Potential-divider method of biasing of JFET.

The resistors R_1 and R_2 form a potential divider across drain power supply V_{DD}.

$$V_2 = \frac{V_{DD}}{R_1 + R_2} \times R_2$$

and

$$V_2 = V_{GS} + I_D R_S$$

or

$$V_{GS} = V_2 - I_D R_S$$

V_2 is smaller than $I_D R_S$ as per design to keep V_{GS} negative for proper biasing. From above equation, we get:

$$I_D = \frac{V_{DD} - V_D}{R_D}$$

and

$$V_{DS} = V_{DD} - I_D (R_D + R_S)$$

4.8 JFET CONNECTIONS

It can be connected in three configurations namely common gate (**CG**), common source (**CS**) and common drain (**CD**).

4.8.1 Common Gate JFET Configuration

Fig. 4.12 shows the circuit of a common gate amplifier. The equivalent circuit of the same is shown in Fig. 4.13.

Field Effect Transistor (FET)

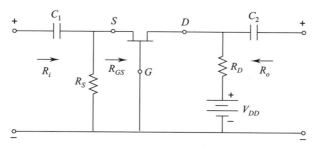

Fig. 4.12 JEET Common gate configuration.

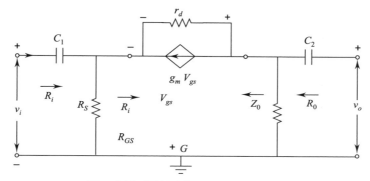

Fig. 4.13 JEET AC equivalent model.

The last JEET configuration to be analyzed in detailed is the common gate configuration of Fig. which parallel the common base configuration employed with BJT transistors.

The Network of intrest is redrawn in Fig. 4.14. The voltage $V' = V_{GS}$. Applying Kirchoff's voltage around the output parameter of the network results in

$$V' - V_{RD} - V_{RD} = 0$$
$$V_{RD} = V' - V_{RD}$$
$$= V' - I_{RD}$$

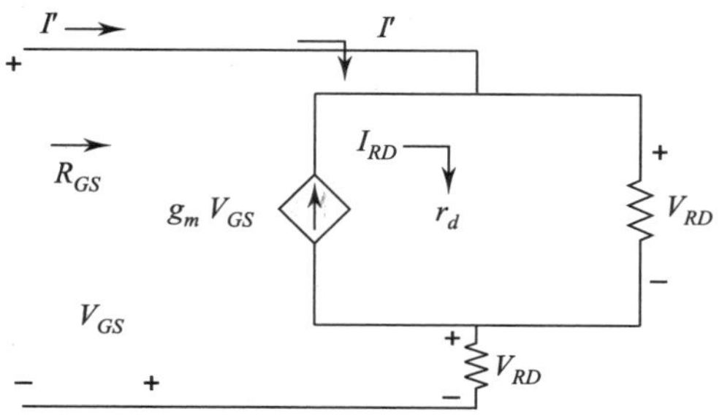

Fig. 4.14 Determining Zi for the Network.

Applying Kirchoff's current law at mode results in

$$I' + g_m V_{GS} = I_{RD}$$

and

$$I' = g_m V_{GS}$$

$$= \frac{V' - I'R_D}{R_D} - g_m V_{GS}$$

or

$$I' = \frac{V'}{R_D} - \frac{I'R_D}{R_D} - g_m[-V']$$

so that

$$I' = \left[1 + \frac{R_D}{r_d}\right]$$

$$= V'\left[\frac{1}{r_d} + g_m\right]$$

and

$$R_{GS} = \frac{V'}{I'} = \frac{\left[1 + \dfrac{R_D}{r_d}\right]}{\left[g_m + \dfrac{1}{r_d}\right]}$$

or

$$\boxed{R_{GS} = \frac{V'}{I'} = \frac{r_d + R_D}{1 + g_m r_d}}$$

$$R_{GS} = \frac{V'}{I'} = \frac{r_d + R_D}{1 + g_m r_d}$$

and

$$R_{in} = R_S \parallel R_{GS}$$

which results in

$$\boxed{R_{in} = R_S \parallel \left[\frac{r_d + R_D}{1 + g_m r_d}\right]}$$

If $r_d \geq 10R_D$, Eq. - Permits the following approximation since $\dfrac{R_D}{V_d} < 1$ and $\dfrac{1}{r_d} \ll g_m$

$$R_{GS} = \frac{\left[1 + \dfrac{R_D}{r_d}\right]}{g_m + \dfrac{1}{r_d}} \cong \frac{1}{g_m}$$

and

$$\boxed{R_i \cong R_S \parallel \frac{1}{g_m}}$$

Z_0 substitutions $V_i = OV$ in figure will 'short-out' the effects of R_S and get V_{GS} to OV. The results is $g_m V_{GS} = 0$, and r_d will be in parallel with R_D. There are

$$\boxed{R_0 = R_D \| r_d}$$

For
$$r_d \geq 10 R_D$$

$$\boxed{R_0 \cong R_D}$$

Figure reveals that

$$v_i = -V_{GS}$$

and
$$v_o = I_D R_D$$

The voltage across r_d is

$$V_{RD} = v_o - v_i$$

and
$$I_{RD} = \frac{v_o - v_i}{r_d}$$

Applying Kirchoff's current law and note b in figure result in

$$I_{RD} + I_D + g_m V_{GS} = 0$$

$$I_D = -I_{RD} - g_m V_{GS}$$

$$= -\left[\frac{v_o - v_i}{r_d}\right] - g_m [-v_i]$$

$$I_D = \frac{v_i - v_o}{r_d} + g_m v_i$$

So that
$$v_o = I_D R_D$$

$$= \left[\frac{v_i - v_o}{r_d} + g_m v_i\right] R_D$$

$$= \frac{v_i - v_o}{r_d} - \frac{v_o R_D}{r_d} + g_m v_i R_D$$

$$v_o \left[1 + \frac{R_D}{r_d}\right] = v_i \left[\frac{R_D}{r_d} + g_m R_D\right]$$

$$A_v = \frac{\frac{R_D}{r_d} + g_m R_D}{1 + \frac{R_D}{r_d}}$$

for $r_d \geq 10 R_D$

$$A_v \cong g_m R_D$$

4.8.2 Common Source JFET Configuration

Common source JFET configuration is shown in Fig. 4.15 and the equivalent circuit of the same is shown in Fig. 4.16.

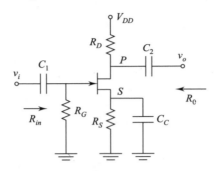

Fig. 4.15 Common Source JFET configuration.

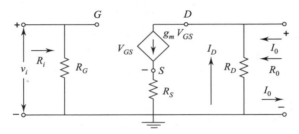

Fig. 4.16 Equivalent circuit.

Z_i Due to open circuit condition between the gate and output network the input remains the following:

$$\boxed{R_i = R_G}$$

Z_0 The output impedence is defined by

$$R_0 = \left.\frac{v_o}{I_0}\right|_{v_i=0}$$

Setting $v_i = OV$ in figure results in the gate terminal being at around potential (OV). The voltage across R_G is then OV and R_G has been effictively "sorted out" of the picture.

Applying Kirchoff's current law results in

$$I_0 + I_D = g_m V_{GS}$$

with
$$V_{GS} = -(I_0 + I_0) R_S$$
so that
$$I_0 + I_D = -g_m (I_0 + I_D) R_S = -g_m I_0 R_S - g_m I_D R_S$$
or
$$I_0 [1 + g_m R_S] = -I_D [1 + g_m R_S]$$
and $I_0 = -I_D$ (the controlled current source gm $V_{GS} = OA$ for the applied conditions)

Field Effect Transistor (FET)

$$v_o = I_D R_D$$
$$v_o = (-I_o) R_D = I_o R_D$$

$$\boxed{R_0 = \frac{v_o}{v_o} = R_D}$$

If R_D is included in the Network the equivalent will appear as shown in Fig. 4.17.

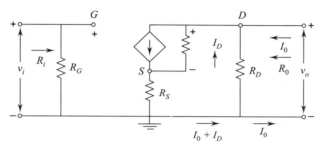

Fig. 4.17

Since
$$R_0 = \frac{v_o}{I_0}\bigg|_{v_i = ov} = -\frac{I_D R_D}{I_0}$$

We short try to find an expression for I_0 in terms of I_D.
Applying Kirchoff's current law we have

$$I_0 = g_m V_{GS} + I_{rd} \frac{v_o + V_{GS}}{r_d} - I_D$$

$$V_{rd} = v_o + V_{GS}$$

and
$$I_0 = g_m V_{GS} + - I_D$$

or
$$I_0 = \left(g_m + \frac{1}{r_d}\right) V_{GS} - \frac{I_d r_d}{r_d} - I_0 \text{ using } V_D = -I_D R_D$$

Now
$$V_{gs} = -(I_D + I_0) R_s$$

So that
$$I_0 = -(I_D + I_0) R_s - \frac{I_D R_D}{r_d} - I_D$$

with the result that

$$I_0 \left[1 + g_m R_s + \frac{R_s}{r_d}\right] = -I_D \left[1 + g_m R_s + \frac{R_s}{r_d} + \frac{R_D}{r_d}\right]$$

or
$$I_0 = \frac{-I_D \left[1 + g_m R_s + \frac{R_s}{r_d} + \frac{R_D}{r_d}\right]}{1 + g_m R_s + \frac{R_s}{r_d}}$$

and
$$R_o = \frac{v_o}{I_0} = \frac{-I_D R_D}{-I_D \left(1 + g_m \frac{R_s}{r_d} + \frac{R_d}{r_d}\right)}$$
$$1 + g_m R_s + \frac{R_s}{r_d}$$

$$R_o = \frac{1 + g_m R_s + \frac{R_s}{r_d}}{\left[1 + g_m R_s + \frac{R_s}{r_d} + \frac{R_D}{r_d}\right]} R_d$$

For $\quad r_d \geq 10\, R_D$
$$R_o = R_D$$

AV - For the network of Fig. 4.17 application of Kirchhoff's voltage law to the input circuits results.
$$v_i - V_{GS} - V_{RS} = 0$$
$$V_{GS} = v_i - I_D R_D$$

The voltage across R_D using Kirchhoff's voltage law is
$$V_{RD} = v_0 - V_{RS}$$

and
$$I' = \frac{V_{RD}}{r_d} = \frac{v_0 - V_{Rs}}{r_d}$$

So that application of Kirchhoff current law results in
$$I_D = g_m\, V_{GS} + \frac{v_0 - V_{Rs}}{r_d}$$

Substituting for V_{GS} from above and substituting for V_0 and V_{RS}, we have
$$I_D = g_m [v_i - I_D R_S] + \frac{(-I_D R_D) - I_D R_s}{r_d}$$

So that
$$I_D \left[1 + g_m R_s + \frac{R_D + R_s}{r_d}\right] = g_m v_i$$

$$I_D = \frac{g_m v_i}{1 + g_m R_s + \frac{R_D + R_s}{r_d}}$$

The output voltage is then

$$v_o = -I_D R_D$$

$$= \frac{g_m R_D v_i}{1 + g_m R_s + \frac{R_D + R_s}{r_d}}$$

$$A_v = \frac{v_o}{v_i}$$

$$= -\frac{g_m R_D}{1 + g_m R_s + \frac{R_D + R_s}{r_d}}$$

Agan if $r_d \geq 10\,(R_D + R_s)$

$$A_V = \frac{v_o}{v_i} = -\frac{g_m R_D}{1 + g_m R_s}$$

4.8.3 Common Drain JFET Configuration

Fig. 4.18 shows the common drain (CD) JFET configuration and the equivalent circuit of the same is shown in Fig. 4.19.

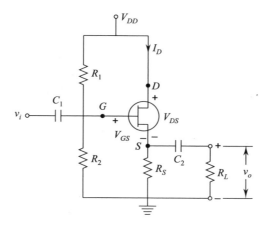

Fig. 4.18 CD JFET Amplifier circuit.

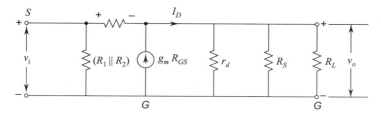

Fig. 4.19 Equivalent circuit for CD amplifier.

We can see that : $V_{GS} = v_i - v_o$ and $R_G = R_1 \| R_2$
and also $\quad v_o = I_D (r_d \| R_S \| R_L)$
or $\quad v_o = g_m V_{GS} (r_d \| R_S \| R_L)$

We can also see that:
$$v_i = V_{GS} + v_o = V_{GS} + V_{GS} g_m (r_d \| R_S \| R_L)$$

We consider that $\quad r_d \gg R_S \| R_L$

∴ Voltage gain, $\quad A_v = \dfrac{v_o}{v_i}$

or
$$A_v = \dfrac{g_m (R_S \| R_L)}{1 + g_m (R_S \| R_L)}$$

We should also note that:
$$R_{in} = R_G = R_1 \| R_2$$

For output resistance, we get :
$$V_{GS} = v_o \times \dfrac{R_{GS}}{(R_{GS} \| R_G) + R_{GS}}$$

We take, $\quad R_{GS} \gg (R_S \| R_G)$, hence, $V_{GS} = v_o$ and $I_D = g_m V_{GS}$

Further, $\quad R_S = \dfrac{v_o}{I_D} = \dfrac{V_{GS}}{g_m V_{GS}} = \dfrac{1}{g_m}$

In fact, r_d is in parallel with $\dfrac{1}{g_m}$,

but $\quad r_d \gg \dfrac{1}{g_m}$.

∴ $\quad R_o = \left(R_S \| \dfrac{1}{g_m} \right)$.

4.9 METAL OXIDE SEMICONDUCTOR FIELD EFFECT TRANSISTOR (MOSFET)

Metal Oxide Semiconductor Field Effect Transistor (**MOSFET**) is a semiconductor device which is similar to JFET with some modifications. The constructional details of N-channel MOSFET are shown in Fig. 4.20. It has only one P-region which is called substrate. A thin layer of metal oxide normally silicon oxide, is deposited over the left side of the channel. Over the oxide layer, a metallic gate is deposited. Since silicon dioxide (SiO_2) is an insulator, hence, gate is insulated from the channel. Due to this fact, MOSFET is also known as Insulated Gate Field Effect Transistor (**IGFET**). Similar to JFET, MOSFET also has three terminals namely source (*S*), gate (*G*) and drain (*D*). Electrons flow from source to drain through *N*-channel.

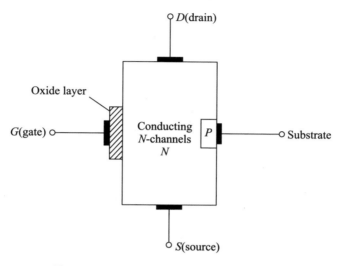

Fig. 4.20 *N*-channel MOSFET construction.

Symbol of *N*-channel MOSFET is shown in Fig. 4.20. Gate looks like a capacitor plate. The thick dark line represents the channel. The top lead is drain and bottom lead is source. The arrow points from substrate to *N*-channel material. Normally, substrate terminal is internally connected, therefore, *N*-channel MOSFET symbol becomes as shown in Fig. 4.21(*b*).

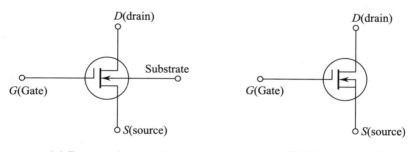

(*a*) Four terminal symbol.   (*b*) Three terminal symbol.

Fig. 4.21 *N*-channel MOSFET symbol and internal connection.

Similarly *P*-channel MOSFET construction is shown in Fig. 4.22. The parts are quite clear except that *N*-type substrate contructs the conducting *P*-channel existing between the source and drain. In this case, holes flow from source to drain the narrow *P*-channel. The *P*-channel MOSFET symbols are shown in Fig. 4.23.

Similar to N-channel MOSFET Fig. 4.23(a) shows P-chnnel MOSFET with four terminals; and Fig. 4.23(b) shows with three terminals wherein substrate is internally connected.

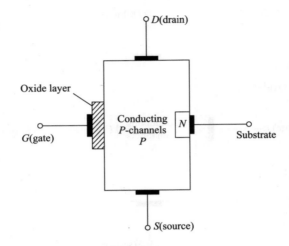

Fig. 4.22 P-channel MOSFET construction.

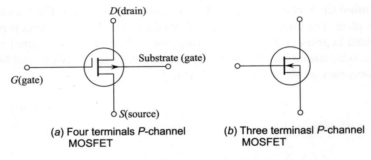

(a) Four terminals P-channel MOSFET

(b) Three terminasl P-channel MOSFET

Fig. 4.23 Symbol of P-channel MOSFET.

It is important to note that the arrow is on the channel and points to the substrate, flow of holes takes place from source to drain through the narrow P-channel.

4.10 MOSFET OPERATION

The gate and the N-channel act like the plates of a capacitor. The metal oxide layer acts as dielectric between two capacitor plates. Consider the circuit of Fig. 4.24. If the gate voltage is changed then, the electric field of the so called capacitor changes. Consequently, the resistance of the N-channel is changed. We can apply either negative or positive voltage on the gate as it is insulated from the N-channel. When negative voltage is applied, the operation is known as Depletion mode. On the other hand, when positive voltage is applied at the gate, the operation is known as

Enhancement mode. Any MOSFET which can be operated, both in depletion and enhancement modes, is designated as DE-MOSFET.

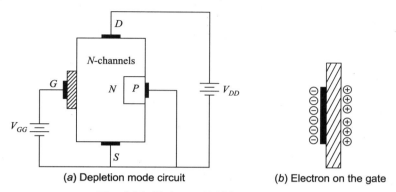

(a) Depletion mode circuit   (b) Electron on the gate

Fig. 4.24 N-channel MOSFET circuit.

4.10.1 Depletion Mode Operation

The circuit used for **depletion mode** operation is shown in Fig. 4.24(a). As the gate is negative, hence, electrons are on gate acting as plate of capacitor. The electrons on the plate repel the free electrons in N-channel. This leaves a layer of positive ions on the part of N-channel next to oxide layer as shown in Fig. 4.24(b). Consequently, N-channel is emptied or depleted of some electrons. Hence, lesser number of free electrons are available for conduction of current through N-channel. It is just like N-channel resistance increase. If the negative voltage on the gate is increased, the current from source to drain is reduced. It can be seen that a change of negative voltage at the gate, changes the resistance of N-channel. This leads to change in current from source to drain.

4.10.2 Enhancement Mode Operation

A circuit for **enhancement mode** operation of MOSFET is shown in Fig. 4.25(a). As explained earlier, the gate and N-channel act as plates and oxide layer as insulator of a capacitor. As the gate is positive in this case, it induces negative change in the N-channel as shown in Fig. 4.25(b).

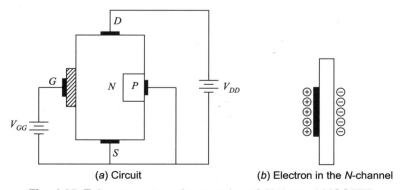

(a) Circuit   (b) Electron in the N-channel

Fig. 4.25 Enhancement mode operation of N-channel MOSFET.

Basically, the negative charges on the plate of capacitor are electrons. Hence, the number of free electrons in the N-channel are increased. Consequently, the conduction of current from source to drain is increased *i.e.*, resistance of N-channel is reduced. Thus, higher the positive voltage on the gate, lower is the current conduction from source to drain. In short, any change of the positive voltage on the gate changes the conductivity of the N-channel. Any increase in positive voltage on the gate enhances the conductivity of N-channel, hence, it is known as Enhancement Mode of operation.

4.11 CHARACTERISTICS OF MOSFET

It has been explained that in an N-channel MOSFET, the gate (positive plate), metal oxide film (dielectric), and substrate (negative plate) form a capacitor. The electric field of this capacitor controls N-channel resistances. The N-channel resistance is dependent on the potential of the gate. If the gate is negative, then the N-channel resistance increases. However, if the gate is positive, the N-channel resistance decreases. Typical drain characteristics is shown in Fig. 4.26 for threshold voltage $V_T = 2$ V when the N-channel MOSFET operation begins. Threshold voltage is the one when no drain current flows.

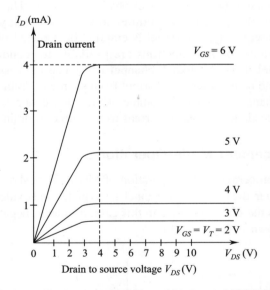

Fig. 4.26 Drain characteristics of N-channel MOSFET.

The transfer characteristics of depletion mode N-channel MOSFET operation is shown in Fig. 4.27. In this case, the gate voltage must be sufficiently negative to ensure that no drain current flows in the OFF condition. Any suitable voltage between this value and zero results in the device being switched ON.

Field Effect Transistor (FET)

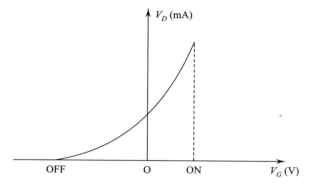

Fig. 4.27 *N*-channel MOSFET (depletion) transfer characteristics.

The transfer characteristics of enhancement mode *N*-channel MOSFET operation is shown in Fig. 4.28. The operation in enhancement mode avoids the need for negative voltage to ensure the OFF operation since this is its normal condition. Again a positive gate voltage causes ON operation to occur.

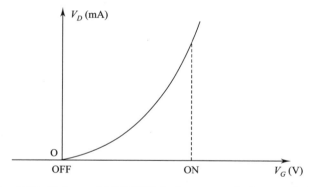

Fig. 4.28 *N*-channel MOSFET (enhancement) characteristics.

Salient points of MOSFET Operation are:

The gate, oxide layer and *N*-channel form a capacitor in the operation of a MOSFET, therefore, the following advantages may be noted :

(*i*) The drain current is controlled by voltage at the gate.

(*ii*) It can be operated with positive and negative gate voltage.

(*iii*) Negligible gate current flows irrespective of gate having negative or positive voltage, hence, the input impedance is very high in the range of several thousand megaohms.

SOLVED EXAMPLES

Example 4.1 The pinch-down voltage of a P-channel junction FET is $V_P = 5$ V and the drain to source saturation current $I_{DSS} = -40$ mA. The value of drain-source voltage V_{DS} is such that the transistor is operating in the saturated region. The drain current is given as $I_D = -15$ mA. Determine the gate-source voltage V_{GS}.

Solution: It is found experimentally that a square-law characteristic closely approximate the drain current in saturation.

$$I_{D\,(sat)} \approx I_{DSS}\left[1 - \frac{V_{GS}}{V_P}\right]^2$$

given that:
$$V_P = 5 \text{ V}$$
$$I_{DSS} = -40 \text{ mA}$$
$$I_D = -15 \text{ mA}$$

$$-15 \text{ mA} = -40 \text{ mA}\left[1 - \frac{V_{GS}}{5}\right]^2$$

or
$$\sqrt{\frac{15}{40}} = 1 - \frac{V_{GS}}{5}$$

∴
$$1 - \frac{V_{GS}}{5} = 0.612$$

$$\frac{V_{GS}}{5} = -0.612 - 1 = 0.3876$$

$$V_{GS} = -1.938 \text{ volt. +ve.}$$

Example 4.2. A JFET amplifier with stabilised biasing circuit shown below. has following parameters:

$V_P = -2$ V, $V_{DSS} = 5$ mA, $R_L = 910$ Ω, $R_F = 2.29$ kΩ, $R_1 = 12$ MΩ, $R_2 = 8.57$ MΩ and $V_{DD} = 24$ V.

Determine the value of drain current I_D at the operation point. Also, varify that FET will operate in pinch-off region.

Solution: Given that drain supply Vol,
$$V_{DD} = 24 \text{ V}$$
Load Resistance, $R_L = 910$ Ω
Source Circuit resistance $R_F = 2.29$ kΩ
Drain source saturation current
$$I_{DSS} = 5 \text{ mA}$$
$$= 5 \times 10^{-3} \text{ A}$$

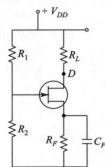

Field Effect Transistor (FET)

Pinch-off voltage $V_P = -2$ V
$$R_1 = 12 \text{ M}\Omega$$
$$R_2 = 8.57 \text{ M}\Omega$$

Now, we know that the gate vol is given by

$$V_G = \frac{V_{DD} \times R_2}{R_1 + R_2} = \frac{24 \times 8.57}{12 + 8.57} = 10 \text{ V}$$

Also, drain current I_D, is given by

$$I_D = I_{DSS}\left[1 - \frac{V_{GS}}{V_P}\right]^2 = 5 \times 10^{-3}\left[\frac{1 - 10 - 2.29 \times 10^3 I_D}{-2}\right]^2$$

or $\qquad I_D = 5 \times 10^{-3} [1 + 5 - 1,145 I_D]^2 = 0.005 [6 - 1,145 I_D]^2$

Solving, we get $I_D = 4.46$ mA.

Gate source voltage, V_{GS} is given by
$$V_{GS} = V_G - I_D R_F = 10 - 2.29 \times 10^3 \times 4.46 \times 10^{-3}$$
or $\qquad V_{GS} = -0.2134$ volt.

Current at operating point is given by

$$I_D = \frac{V_G - V_{GS}}{R_F} = \frac{10 - 0.2134}{2.29 \times 10^3} = 4.27 \text{ mA}.$$

It may be noted that the value of I_D at operating point is almost equal to previously computed value of I_D. Therefore, we can say that the FET is operating in pink-off region.

Example 4.3 A JFET amplifier has $g_m = 2.5$ mA/V and $r_d = 500$ kW. The load resistance is 10 kW. Find the value of voltage gain.

Solution: Given that
$$g_m = 2.5 \text{ mA/V} = 2.5 \times 10^{-3} \text{ A/V} ;$$
$$r_d = 500 \text{ k}\Omega - 8R_D = 10 \text{ k}\Omega.$$

We know that the ac equivalent resistance is given by

$$r_L = \frac{R_D \times r_d}{R_D + r_d} = \frac{10 \times 500}{10 + 500} \text{ k}\Omega = 9.8 \text{ k}\Omega$$

$$= 9.8 \times 10^3 \text{ }\Omega$$

Now voltage gain $\qquad A_v = -g_m \times r_L = -(2.5 \times 10^{-3}) \times (9.8 \times 10^3)$
$$A_v = 24.5.$$

Example 4.4 An FET follows the following relations:

$$I_D = I_{DSS}\left[1 - \frac{V_{GS}}{V_P}\right]$$

I_{DSS} = 8.4 mA, V_P = – 3 V. What is the value of I_D for V_{GS} = – 1.5 V ? Find *gm* at this point.

Solution: Drain-source saturation current,
$$I_{DSS} = 8.4 \text{ mA}$$
Pinch-off voltage, $\quad V_P = -3 \text{ V}$
Gate-source voltage, $\quad V_{GS} = -1.5 \text{ V}$
We know that the drain current is

$$I_D = I_{DSS}\left[1 - \frac{V_{GS}}{V_P}\right]^2$$

Substituting all the values, $I_D = 8.4\left[1 - \frac{-1.5}{-3}\right]^2 = 2.1$ mA.

Transconductance for $V_{GS} = 0$ is given by

$$g_{m_0} = \frac{-2I_{DSS}}{V_P} = \frac{-2 \times 8.4}{-3} = 5.6 \text{ mA/V or 5.6 mS.}$$

Transconductance, $\quad g_m = g_{m_0}\left[1 - \frac{V_{GS}}{V_P}\right] = 5.6\left[1 - \frac{-1.5}{-3}\right] = 2.8$ mS.

Example 4.5 An *N*-channel JFET has a pinch off voltage of – 4.5 V and I_{DSS} = 9 mA. At what value of V_{GS} will I_{DS} equal to 3 mA ? What is its g_m at this I_{DS} ?

Solution: Pinch-off voltage
$$V_P = -4.5 \text{ V}$$
Drain-source saturation current
$$I_{DSS} = 9 \text{ mA} = 9 \times 10^{-3} \text{ A}$$
Drain source current
$$I_{DS} = 3 \text{ mA} = 3 \times 10^{-3} \text{ A.}$$
From Shockleys equation

$$V_{GS} = V_P\left[1 - \sqrt{\frac{I_{DS}}{I_{DSS}}}\right] = -4.5 \times \left[1 - \sqrt{\frac{3 \times 10^{-3}}{9 \times 10^{-3}}}\right] = 1.902 \text{ V.}$$

Transconductance g_m for $I_{DS} = 3$ mA for which
$$V_{GS} = -1.902 \text{ V}$$

$$g_m = \frac{-2I_{DSS}}{V_P}\left[1 - \frac{V_{GS}}{V_P}\right] = \frac{-2 \times 9 \times 10^{-3}}{-4.5}\left(1 - \frac{-1.902}{-4.5}\right)$$

$$g_m = 2.31 \text{ mA/V} = 2.31 \text{ mS}.$$

Example 4.6 A JFET has $V_P = -4.5$ V, $I_{DSS} = 10$ mA and $I_{DS} = 2.5$ mA. Determine the transconductance.

Solution: Drain-source saturation current

$$I_{DSS} = 10 \text{ mA}$$

Pinch-off voltage, $\quad V_P = -4.5$ V

Drain-source current $\quad I_{DS} = 2.5$ mA

From Shockleys equation, drain-source current,

$$I_{DS} = I_{DSS}\left[1 - \frac{V_{GS}}{V_P}\right]^2$$

or

$$V_{GS} = V_P\left[1 - \sqrt{\frac{I_{DS}}{I_{DSS}}}\right] = -4.5\left[1 - \sqrt{\frac{2.5}{10}}\right]$$

$$V_{GS} = 22.5 \text{ V}$$

Transconductance,

$$g_m = -\frac{2I_{DSS}}{V_P}\left[1 - \frac{V_{GS}}{V_P}\right] = \frac{-2 \times 10 \times 10^{-3}}{-4.5}\left[1 - \frac{-2.25}{4.5}\right] = 2.22 \text{ mA/V}.$$

Example 4.7 An n-channel JFET has $I_{DSS} = 10$ mA and $V_P = -9$ V. Determine the minimum value of V_{DS} for pinch-off region and drain current I_D for $V_{GS} = -2$ V in pinch-off-regioin.

Solutioin: Pinch-off voltage,

$$V_P = -4 \text{ V}$$

Gate-source voltage, $\quad V_{GS} = -2$ V

Drain-Source Saturation current,

$$I_{DSS} = 10 \text{ mA} = 10 \times 10^{-3} \text{ A}$$

Drain current,

$$I_D = I_{DSS}\left[1 - \frac{V_{GS}}{V_P}\right]^2$$

$$= 10 \times 10^{-3}\left[1 - \frac{-2}{-4}\right]^2 = 2.5 \text{ mA}.$$

The minimum value of V_{DS} for pinch-off region is equal to V_P. Thus, the minimum value of V_{DS}.

$$V_{DS \text{ (min)}} = V_P = -4 \text{ V}.$$

Example 4.8 A data sheet gives these JFET values
$$I_{DSS} = 20 \text{ mA} \quad \text{and} \quad V_p = 5 \text{ V}.$$
What is the maximum drain current ? What is the gate-source cut-off voltage ?

Solution: For any gate voltage, the drain current has to be in this range,
$$0 < I_D < 20 \text{ mA}$$
When the gate voltage is zero, the drain current has its maximum value of $I_D = 20$ mA.

The gate-source voltage has the same magnitude as the pinch off voltage but the opposite sign. Since the pinch-off voltage is 5 V,
$$V_{GS \text{ (off)}} = -5 \text{ V}.$$

Example 4.9 Suppose a JFET has $I_{DSS} = 7$ mA and $V_{GS \text{ (off)}} = -3$ V. Calculate the drain current for a gate-source voltage of -1 V.

Solution:

With equation (i), you can work out the k factor as follows:
$$K = \left(1 - \frac{V_{GS}}{V_{GS \text{ (off)}}}\right)^2 \qquad \ldots(i)$$

$$K = \left(1 - \frac{1 \text{ V}}{3 \text{ V}}\right)^2 = (0.667)^2 = 0.445$$

Now, multiply the K-factor by I_{DSS} to get the drain current.
$$I_D = 0.445 \ (7 \text{ mA}) = 3.12 \text{ mA}.$$

Example 4.10 In the given figure, the resistor is changed to 3.6 kW. If $V_{GS} = 0$, what is the drain-source voltage?

Solution: Assume the JFET act as a current source. Since the ground voltage is zero, the drain current is at its maximum value of 10 mA.

Therefore, the drain source voltage is
$$V_{DS} = 10 \text{ V} - (10 \text{ mA}) \ (3.6 \text{ k}\Omega) = -26 \text{ V}$$

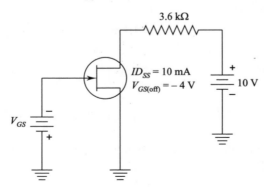

Field Effect Transistor (FET)

Impossible ! The drain voltage current be (–) ve. We have an absurd result, which means the JFET cannot be operating in the current-source region. It must be operating in the ohmic region.

Here is what to do next. Since the JFET is operating in the ohmic region, we need to calculate the value of R_{DS}. It equals the pinch-off voltage divided by the maximum drain current.

$$R_{DS} = \frac{4\text{ V}}{10\text{ mA}} = 400\ \Omega.$$

The equivalent circuit for the drain current is given below. The drain-source voltage can be calculated as follows:

$$V_{DS} = \frac{10\text{ V}}{4\text{ k}\Omega}(400\ \Omega) = 1\text{ V}$$

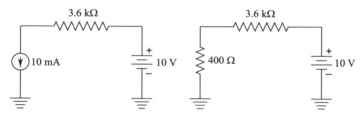

Equivalent circuit

Example 4.11 What is the drain-source voltage in Example 4.10 for $V_{GS} = -2.2$ V ?

Solution: Since V_{GS} has changed from 0 to – 2.2 V, there is less drain current and it is possible that the JFET no longer operates in the ohmic region. Here is how to proceed. Assume the JFET is operating as a current source. JST, gets the K-factor and the drain current as follows :

$$K = \left(1 - \frac{2.2\text{ V}}{4\text{ V}}\right)^2 = (0.45)^2 = 0.203.$$

and
$$I_D = 0.203\ (10\text{ mA}) = 2.03\text{ mA}.$$

Second, the drain-source voltage is

$$V_{DS} = 10\text{ V} - (2.03\text{ mA})(3.6\text{ k}\Omega) = 2.69\text{ V}$$

Third, calculate the proportional pinch-off vol:

$$V_P = (2.03\text{ mA})(400\ \Omega) = 0.812\text{ V}.$$

This voltage separates the ohmic region and the active region when $V_{GS} = -2.2$ V. Since a V_{DS} of 2.69 V is greater than a $V'P$ of 0.812 V, the JFET is operating as a current source. This agrees with the original assumption. Therefore, the final answer is

$$V_{DS} = 2.69\text{ V}.$$

Example 4.12 In the given figure the resistor is changed to 4.7 kΩ. If $V_{GS} = 0$, what is the drain-source voltage?

Solution: Assume the MOSFET acts like a current source. Since the gate vol is zero, the drain current is 10 mA.

Therefore, the drain-source voltage is

$$V_{DS} = 20 \text{ V} - (10 \text{ mA})(4.7 \text{ k}\Omega) = -27 \text{ V}$$

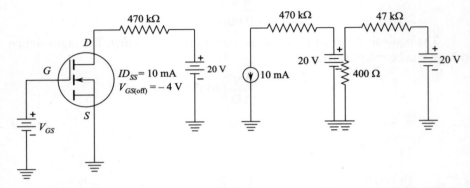

Impossible! the drain voltage cannot be (−) ve. We have an absurd result, which means the MOSFET cannot be operating in the active region. It must be operating in the ohmic region.

Here is what to do next. Since the MOSFET is operating in the ohmic region, we need to calculate the value of R_{DS}. It equals the pinch-off vol divided by the maximum drain current.

$$R_{DS} = \frac{4 \text{ V}}{10 \text{ mA}} = 400 \text{ }\Omega.$$

The MOSFET acts like a resistance of 400 W. The total resistance in the drain circuit is the sum of 400 W and 4.7 kW.

Therefore, the drain-source vol is

$$V_{DS} = \frac{20 \text{ V}}{5.1 \text{ k}\Omega}(400 \text{ }\Omega) = 1.57 \text{ V}.$$

Example 4.13 In the given figure shown in example 4.12. What is the drain-source vol when $V_{GS} = +1 \text{ V}$?

Solution: Assume the MOSFET is operating as a current source. First, get the K factor and the drain current as follows :

$$K = \left(1 - \frac{+1 \text{ V}}{-4 \text{ V}}\right)^2 = (1.25)^2 = 1.56$$

and

$$I_D = 1.56 (10 \text{ mA}) = 15.6 \text{ mA}$$

Field Effect Transistor (FET)

Second, the drain-source vol is
$$V_{DS} = 20 \text{ V} - (15.6 \text{ mA})(470 \text{ Ω}) = 12.7 \text{ V}$$
Third, calculate the proportional pinch-off vol :
$$V'_P = (15.6 \text{ mA})(400 \text{ Ω}) = 6.24 \text{ V}$$

Since V_{DS} is greater than V'_P, the MOSFET is operating as a current source.

Example 4.14 In the given figure, the drain resistor increases to 36 kΩ. What is the drain-source vol when V_{GS} is 5 V ?

Solution:

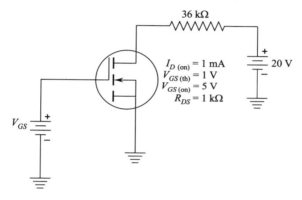

Assume the MOSFET acts as a current source. Since the gate voltage is + 5 V, the drain current is 1 mA. Therefore, the drain source voltage is
$$V_{DS} = 20 \text{ V} - (1 \text{ mA})(36 \text{ kΩ}) = -16 \text{ V}.$$

Impossible ! the drain vol. cannot be –ve. We have an absurd result, which means the assumption about the current source is incorrect. The MOSFET cannot be operating in the current source region. It must be operating in the ohmic region.

The MOSFET acts as a resistance of 1 kΩ. The total resistance in the drain circuit is the sum of 1 and 36 kΩ. Therefore, we can calculate the drain-source vol to like this :
$$V_{DS} = \frac{20 \text{ V}}{1 \text{ kΩ} + 36 \text{ kΩ}}(1 \text{ kΩ}) = 0.54 \text{ V}.$$

Example 4.15 In above figure, what is the drain source voltage when $V_{GS} = 3 \text{ V}$?

Solution: Assume the MOSFET is operating as a current source. First, get the K factor by substituting the given equation (i).

$$K = \left(\frac{V_{GS} - V_{GS\,(th)}}{V_{GS\,(on)} - V_{GS\,(th)}}\right)^2 \qquad \ldots(i)$$

$$K = \left(\frac{3\text{ V} - 1\text{ V}}{5\text{ V} - 1\text{ V}}\right)^2 = (0.5)^2 = 0.25$$

and $I_D = 0.25\ (1\text{ mA}) = 0.25\text{ mA}.$

Second, the drain-source vol is

$$V_{DS} = 20\text{ V} - (0.25\text{ mA})\ (3.6\text{ k}\Omega) = 19.1\text{ V}.$$

Example 4.16 Mathematical derivation

In the enhancement mode MOSFET, the basic equation of drain current was given as

$$I_D = K\ (V_{GS} - V_{GS\ (th)})^2 \qquad \ldots(i)$$

Solution: The derivation of this basic formula is given in engineering textbooks on FETs. Here we want to show how this equation is rearranged into the more useful form

$$I_D = K\ I_{D\ (on)} \qquad \ldots(ii)$$

where,

$$K = \left(\frac{V_{GS} - V_{GS\ (th)}}{V_{GS\ (on)} - V_{GS\ (th)}}\right)^2 \qquad \ldots(iii)$$

To begin, substitute $I_{D\ (on)}$ and $V_{GS\ (on)}$ into equation (*i*) to get

$$I_{D\ (on)} = K\ (V_{GS\ (on)} - V_{GS\ (th)})^2$$

Solve for K to get

$$K = \frac{I_{D(on)}}{\left(V_{GS\ (on)} - V_{GS\ (th)}\right)^2}$$

Substitute this K into equation (*i*) to get

$$I_D = \frac{I_{D(on)}}{\left(V_{GS\ (on)} - V_{GS\ (th)}\right)^2}\left(V_{GS} - V_{GS\ (th)}\right)^2$$

Now, define

$$K = \left(\frac{V_{GS} - V_{GS\ (th)}}{V_{GS\ (on)} - V_{GS\ (th)}}\right)^2$$

which means $I_D = K\ I_{D\ (on)}.$

Example 4.17 A 2 M 5457 has $I_{DSS} = 5$ mA and $g_m 0 = 5000$ m mho. What is the value of I_D for $V_{GS} = -1$ V? What is the gm for this drain current?

Solution: From width equation (*i*) to get an accurate value of $V_{GS\ (off)}$

$$V_{GS\ (off)} = \frac{-2\ I_{DSS}}{g_{m_0}} \qquad \ldots(i)$$

$$= \frac{-2\ (5\text{ mA})}{5000\ \mu\text{ mho}} = -2\text{ V}$$

To get the drain current, first calculate the K factor with equation (ii)

$$K = \left(1 - \frac{V_{GS}}{V_{GS(off)}}\right)^2 \qquad ...(ii)$$

$$K = \left(1 - \frac{1}{2}\right)^2 = (0.5)^2 = 0.25.$$

Then $I_D = 0.25\,(t\text{ mA}) = 1.25\text{ mA}$

Next use equation (iii) to calculate g_m at $V_{GS} = -1$ V:

$$g_m = g_{m_0}\left(1 - \frac{V_{GS}}{V_{GS(off)}}\right) \qquad ...(iii)$$

$$= (5000\ \mu\text{ mho})\left(1 - \frac{1\text{ V}}{2\text{ V}}\right) = 2500\ \mu\text{ mho}.$$

As you see, g_m is 2500 m mho when I_D is 1.25 mA.

Example 4.18 If a JFET has $g_m = 2500$ m mho, what is the ac drain current for $V_{GS} = 1$ mV ? Compare this to a tripolar transistor width a g_m of 50,000 μ mho.

Solution: In the given figure, the current source has a value of $g_m V_{GS}$. Therefore, an ac input of 1 mV produces

$$id = (25000\ \mu\text{ mho})\,(1\text{ mV}) = 2.5\ \mu\text{A}.$$

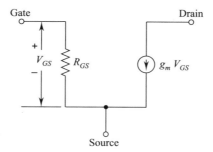

The same ac input voltage to a tripolar transistor would produce an ac collector current of

$$ic = (50{,}000\ \mu\text{ mho})\,(1\text{ mV}) = 50\ \mu\text{A}.$$

This bipolar output current is 20 times greater than the JFET output curent. Given the same load resistances, a bipolar amp'r would produce 20 times more output vol than the JFET.

Example 4.19 If $g_m = 2500$ m mho for the JFET of given circuit, what is the ac output voltage ?

Solution: The ac drain resistance is

$$r_d = 3.6 \text{ k}\Omega \parallel 10 \text{ k}\Omega = 2.65 \text{ k}\Omega$$

The voltage gain is $\quad A = (2500 \text{ } \mu \text{ mho}) (2.65 \text{ k}\Omega) = 6.63$

The input impedance of the amplitude is

$$Z_{in} = 1 \text{ M}\Omega.$$

We are ignoring the R_{GS} of the JFET because it is usually in the hundreds of mega ohm.

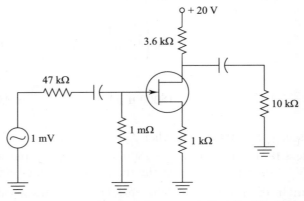

The generator has an internal resistance of 47 kΩ. Therefore, some of the signal voltage is dropped across this 47 kΩ. But not much, the ac voltage at the gate is found with ohm's law.

$$V_{in} = \frac{1 \text{ mV}}{47 \text{ k}\Omega + 1 \text{ M}\Omega}(1 \text{ M}\Omega) = 0.955 \text{ mV}.$$

The ac output voltage equals the V gain time the input voltage

$$V_{out} = 6.63 \text{ } (0.955 \text{ mV}) = 6.33 \text{ mV}.$$

Example 4.20 If $g_m = 2500$ m mho for the source follows of in figure, what is the ac output voltage ?

Solution:

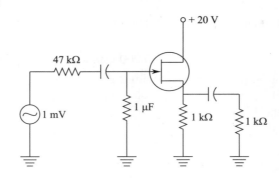

Field Effect Transistor (FET)

The input voltage drives the gate, and the output voltage appears at the source. The ac source resistance is

$$r_s = 1 \text{ k}\Omega \parallel 1 \text{ k}\Omega = 500 \text{ }\Omega$$

Width equation, $A = \dfrac{g_m r_s}{1 + g_m r_s}$, the V gain is

$$A = \dfrac{(2500 \text{ }\mu\text{ mho})(500 \text{ }\Omega)}{1 + (2500 \text{ }\mu\text{ mho})(500 \text{ }\Omega)} = 0.556$$

The input impedance of the source full is

$$Z_{in} = 10 \text{ M}\Omega.$$

The generator has an internal resistance of 47 kΩ. Therefore, almost none of the ac voltage is dropped across the generator resistance :

$$v_i = \dfrac{1 \text{ mV}}{47 \text{ K} + 10 \text{ M}\Omega}(10 \text{ M}\Omega) = 0.995 \text{ mV}$$

The ac output voltage equals the V gain times the input voltage

$$v_o = (0.556)(0.995 \text{ mV}) = 0.553 \text{ mV}.$$

Example 4.21 JFET shunt switch like in the figure has $R_D = 10$ kΩ, $I_{DSS} = 10$ mA and $V_{GS \text{(off)}} = -2$ V. If $v_i = 10$ mV peak to peak, what does v_o equal when $V_{GS} = 0$? When $V_{GS} = -5$ V ?

Solution: Calculate the ideal value of R_{DS} as follows:

$$R_{DS} = \dfrac{2 \text{ V}}{10 \text{ mA}} = 200 \text{ }\Omega$$

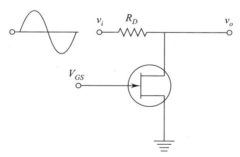

When $V_{GS} = 0$, the circuit acts like the equivalent circuit of given circuit width ohm's law.

$$v_o = \dfrac{10 \text{ mV}}{10 \text{ k}\Omega + 200 \text{ }\Omega}(200 \text{ }\Omega)$$

$$= 0.196 \text{ mV}.$$

Thus, the equivalent circuit is as follows:

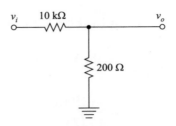

When $V_{GS} = -5$ V, the JFET is like an open circuit.

In Figure visualise the 200 V increasing to input. You can see that

$$v_o = v_i = 10 \text{ mV}.$$

SUMMARY

1. **JFET Basics:** The junction field effect transistor (JFET) has source gate and drain terminals. It has two built in diodes – the gate source-diode and the gate-drain diode. These diodes will conduct if these are forward biased with more than 0.7 V. Both the gate-source diode and gate-drain diode are reverse biased for normal operation. It has input resistance which approaches infinity, but it has less voltage gain than a bipolar transistor. Some additional details are as follows:

   (*i*) JFET can be used as a voltage-controlled resistor as it has a unique sensitivity of the drain-to-source impedance to the gate-to-source voltage.

   (*ii*) The maximum current I_{DSS} occurs when $V_{GS} = 0$ V; and the minimum current occurs at pinch-off defined by $V_{GS} = V_P$.

   (*iii*) The relationship between the drain current and gate to source voltage is non-linear defined by Shockley's equation.

2. **Input and Transfer Characteristics of JFET:** The drain characteristics are similar to those of a BJT, except that V_{GS} is the controlling input rather than I_B. The transconductance characteristics is the plot of drain current versus gate voltage. The curve is non-linear, part of a parabola, and also called a square law curve.

3. **MOSFET Basics:** The metal oxide semiconductor field effect transistor (MOSFET) has source, gate and drain terminals. The gate is electrically insulated from the channel. Due to this, the dc input resistance is even higher than that of a JFET. Some additional details are as follows:

   (*i*) MOSFETs should always be handled with additional conduct the static electricity that exists in places we might least expect. Any shorting mechanism between the leads of the device should not be removed until it is installed.

(ii) A complementary MOSFET (C_{MOS}) device uses a unique combination of a P-channel and an N-channel MOSFET with single set of external leads. It has very high input impedance, fast switching speeds, and low operating power levels.

4. **Depletion Mode MOSFET:** Normally, the depletion-mode MOSFET is ON when V_{GS} is zero. It has drain curves and equivalent circuits similar to JFET except that MOSFET can operate with positive as well as negative gate voltages.

5. **Enhancement Mode MOSFET:** It is normally OFF when the gate voltage is zero. A positive sufficient gate voltage forms it ON. The voltage at which this turns ON is the threshold voltage. It can act as a current source or as a resistor.

   The transfer characteristics of an enhancement type MOSFET are not defined by Shockley's equation but rather by a non-linear equation controlled by the gate-to source voltage, the threshold voltage, and a constant k defined by the device used. The plot of I_D versus V_{GS} rises exponentially with increasing values of V_{GS}.

6. The arrow in the symbol of N-channel JEFTs or MOSFETs join into the centre of symbol, whereas those of a P-channel device will always point out of the centre of the symbol.

7. **Important Equations:**
   (i) Gate Cut-off and Pinch-off.
   $$V_{GS}(\text{off}) = -V_P$$
   (ii) Drain – Source Resistance.
   $$R_{DS} = \frac{V_P}{I_{DSS}}$$
   (iii) Drain Current as a function of Gate voltage:
   $$I_D = k \, I_{DSS}$$
   where
   $$k = \left(1 - \frac{V_{GS}}{V_{GS\,\text{off}}}\right)^2$$
   k value is between 0 and 1.
   (iv) Proportional Pinch-off
   $$V_P' = I_D \, R_{DS}$$
   If V_{DS} is greater than V_P', the JFET acts as a current source; and if V_{DS} is less than V_P', the JFET acts like a resistor.

   (v) Enhancement – Mode Drain Current for MOSFET.
   $$I_D = k \, I_{DOH}$$
   Where
   $$k = \left(\frac{V_{GS} - V_{GS\,(th)}}{V_{GS\,(on)} - V_{GS\,(th)}}\right)^2$$
   for $V_{GS} > V_{GS\,(th)}$

(vi) AC Drain Resistance

$$r_d = \frac{\Delta V_{DS}}{\Delta I_D} \text{ (from drain-to-source)}$$

(vii) Amplification Factor:

$$\mu = \frac{\Delta V_{DS}}{\Delta V_{GS}}.$$

EXERCISES

4.1 Explain basic construction of a JFET with illustrative diagrams.

4.2 Describe principle of working of a JFET with illustrative diagrams.

4.3 What is concept of pinch – off? Explain with diagrams.

4.4 Explain maximum drain saturation current. Where is it applicable?

4.5 Describe input and transfer characteristics of JFET and characteristic equation.

4.6 Give CG, CS and CD configurations of JFET and describe their important aspects.

4.7 Describe fixed biasing of JFET amplifier.

4.8 Describe self-biasing of JFET amplifier.

4.9 Explain basic construction of a JFET with illustrative diagrams.

4.10 What is depletion type MOSFET? Explain with diagrams.

4.11 Explain enhancement type MOSFET with diagrams.

4.12 Describe operation of MOSFETs with diagram. Give its characteristics.

4.13 The data sheet of a certain JFET indicates that I_{DSS} is equal to 15 mA and $V_{GS\,(off)}$ is equal to – 5 volt. Find the drain current for V_{GS} equal to 0 volt, – 1 volt and – 4 volt.

4.14 A JFET has parameters of $V_{GS\,(off)}$ equal to – 20 volt and I_{DSS} equal to 12 mA. Plot the trans conductance curve for the device using V_{GS} values of 0 V, – 10 V, – 15 V and – 20 V.

4.15 A 2N5486 JFET has values of $V_{GS\,(off)}$ equal to – 2 V to – 6 V and I_{DSS} is equal to 8 mA to 20 mA. Plot the minimum and maximum trans conductance curves for the device.

4.16 The following information is included on the data sheet for an N-channel JFET:

$$I_{DSS} = 20 \text{ mA}, V_P = -8 \text{ V, and } g_{m_0} = 5000 \text{ }\mu\text{s}$$

Find the values of the drain current and trans conductance at $V_{GS} = -4$ V.

4.17 The data sheet for a certain enhancement type MOSFET revals that $I_{D\,(on)} = 10$ mA at $V_{GS} = -12$ V and $V_{GS\,(Th)} = -3$ V. Is this device P-channel or N-channel? Find the value of I_D, when $V_{GS} = -6$ V.

4.18 Sketch the transfer curve defined by $I_{DSS} = 12$ mA and $V_P = -6$ V using Shockley's equation.

4.19 Sketch transfer characteristics for n-channel depletion-type MOSFET with $I_{DSS} = 10$ mA and $V_P = -4$ V.

4.20 A data sheet gives these JFET values:

$$I_{DSS} = 20 \text{ mA and } V_P = 5 \text{ V.}$$

Find the dc resistance of the JFET in ohmic region.

4.21 In the figure given below, what is the drain – source voltage when V_{GS} is zero.

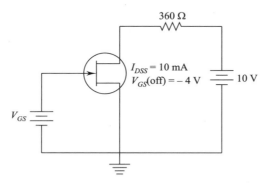

4.22 An n-channel JFET has $I_{DSS} = 8$ mA and $V_P = -5$ V. Find the minimum value of V_{DS} for pinch – off region and the drain current I_{DS} for $V_{GS} = -2$ V in the pinch – off region.

5

OPERATIONAL AMPLIFIER (OP-AMP)

5.1 INTRODUCTION

Operational Amplifier is abbreviated as Op-Amp. Op-Amp is an amplifier which could be easily modified by external circuitry to perform mathematical operations – addition, scaling, integration, *etc*. The advance of solid-state technology has made Op-amp highly reliable, miniaturized, and consistently predictable in performance. Op-amps are building blocks in basic amplification, signal conditioning, active filters, function generators and switching circuits.

5.2 OP-AMP INTEGRATED CIRCUIT

Op-amps are now available in hundred of types. A very good all round performer is the popular LF411 ("411" for short), originally introduced by National Semiconductor. It is packed similar to all op-amps and it looks as shown in Fig. 5.1.

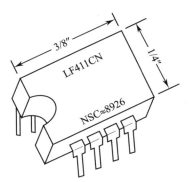

Fig. 5.1 Op-amp integrated circuit.

An op-amp IC 411 is a piece of silicon containing 24 transistors (21 BJTs, 3FETs, 11 resistors and 1 capacitor). The pin configurations are shown in Fig. 5.2. The dot in the corner or notch at the end of the package, identifies the end from which to begin counting the pin numbers. As with most electronic IC packages, pins are run counted clockwise, view counted from the top. The "**offnull**" **terminals** (also known as "**balance**" or "**trim**") have to do with correcting externally the asymmetries that are unavoidable when making op-amp.

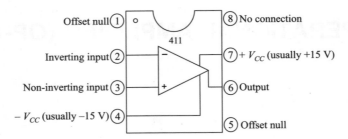

Fig. 5.2 Op-amp 411 pin configurations (top-view).

5.3 OP-AMP SYMBOL

Basic form of operational amplifier is shown in Fig. 5.3. It shows the complete triangular schematic symbol showing the pin connections to different points. Pin 7 connects to $+V_{CC}$ and pin 4 connects to $-V_{CC}$. Pin 6 connects to the Op-amp output. Pin 2 and Pin 3 connect to the Op-amp inputs. Pin 2 is inverting (−) input and pin 3 is non-inverting (+) input.

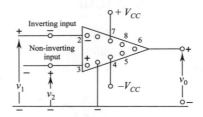

Fig. 5.3 Basic operational amplifier.

Figure 5.4 shows the simplified symbol of Op-amps. This symbol is in most of the representation in various circuits.

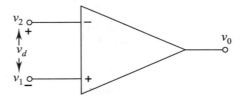

Fig. 5.4 Simplified symbol of Op-amp.

The Fig. 5.4 gives the difference, $v_d = v_1 - v_2$, between two input signals, exhibiting the **open loop gain**

$$A_{OL} = \frac{v_0}{v_d} \qquad \ldots(5.1)$$

Terminal 1, labelled with minus sign, is inverting input; signal v_1 is amplified in magnitude and appears phase-inverted at the output. Conversely, terminal 2, labelled with plus sign, is the non-inverting input; output due to v_2 is phase preserved. In magnitude, open-loop voltage gain in Op-amp ranges from 10^4 to 10^7. The maximum magnitude of output voltage from an Op-amp is called saturation voltage; this voltage is approximately 2 V smaller than power supply voltage. In other words, the amplifier is linear over the range:

$$-(V_{CC} - 2) < v_0 < (V_{CC} - 2)$$

5.4 CONCEPT OF IDEAL OP-AMP

The common **IC Op-amp** has a very high gain. Op-amps are used in analog linear amplification systems and digital logic systems. The properties common to all Op-amps are as follows:

(*i*) an inverting input.

(*ii*) a non-inverting input.

(*iii*) a high-input impedance, normally assumed infinite at both inputs.

(*iv*) a low output impedance.

(*v*) a large voltage gain when operating without feedback, typically 10^5.

(*vi*) voltage gain remains constant over a wide frequency range.

(*vii*) almost no drift due to ambient temperature change, hence, the direct voltage output is zero when there is no input signal.

(*viii*) good stability.

5.5 INVERTING AMPLIFIER

Figure 5.5 shows a basic **inverting amplifier** circuit having two resistors R_1 and R_f to an Op-amp. Normally, the power supply connections to Op-amp are not shown.

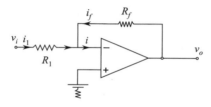

Fig. 5.5 Inverting amplifier.

The open-loop gain of the Op-amp is taken as A. Hence, the output voltage $v_0 = Av_i$.

$$\therefore \qquad v_i - v = i_i R_1$$

Now, if input impedence of Op-amp is very high; then $i \cong 0$.

$$\therefore \qquad i_i = -i_f$$

We also know that:
$$i_f = \frac{v_i - v}{R_1}$$

$$\therefore \qquad \frac{v_i - v}{R_1} = -\frac{v_0 - v}{R_f} = \frac{v - v_0}{R_f}$$

If the input is exactly out of phase with the input voltage, the Op-amp being in its inverting mode, then we get:

$$v_0 = -A v$$

or
$$v = -\frac{v_0}{A}$$

$$\therefore \qquad \frac{v_1 + \dfrac{v_0}{A}}{R_1} = \frac{-\dfrac{v_0}{A} - v_0}{R_f}$$

or
$$v_i + \frac{v_0}{A} = -\frac{v_0}{A} \times \frac{R_1}{R_f} - v_0 \frac{R_1}{R_f}$$

Normally, R_1 and R_f are of the same range of resistance say $R_1 = 100$ k and $R_f = 1$ M, and $A = 10^5$.

$$\therefore \qquad \frac{v_0}{A} \text{ and } \frac{v_0}{A} \times \frac{R_1}{R_f} \text{ can be neglected.}$$

$$\therefore \qquad v_1 = -v_0 \frac{R_1}{R_f} \qquad \qquad ...(5.2)$$

$$\therefore \qquad \text{Overall gain,} \quad \boxed{A_v = -\frac{R_f}{R_1}} \qquad ...(5.3)$$

Normally, in practice, the non-inverting input is earthed through R_2 which minimizes the worst effects of the offset voltage and thermal drift. The offset voltage is the voltage difference between the Op-amp input terminals required to bring the output to zero. Further, the output often includes a resistance of about 50-200 Ω in order to give protection in the event of load being short circuited.

5.6 NON-INVERTING AMPLIFIER

Non-inverting amplifier circuit is shown in Fig. 5.6. It shows two common forms which are identical electrically, but the conversion from one diagram layout to the other can give difficulty.

Due to the very high input resistance, the input current is negligible, therefore, the voltage drop across R_2 is negligible.

$$\therefore \quad v_i = v$$

Using form of Fig. 5.6(b), it can be seen that:

$$v = \frac{R_1}{R_1 + R_f} v_0$$

$$\therefore \quad v_1 = \frac{R_1}{R_1 + R_f} v_0$$

$$\therefore \quad A_v = \frac{v_0}{v_i} = 1 + \frac{R_f}{R_1} \qquad \ldots(5.4)$$

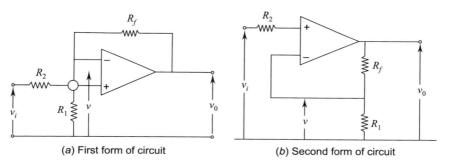

(a) First form of circuit (b) Second form of circuit

Fig. 5.6 Non-inverting amplifier circuit.

5.7 UNITY GAIN OR VOLTAGE FOLLOWER AMPLIFIER

The **unity gain** amplifier circuit is shown in Fig. 5.7. It has voltage gain of 1, and the output is in phase with the input. It also has an extremely high input impedance, leading to use as an intermediate-stage (buffer) amplifier to prevent a small load impedance from loading the input.

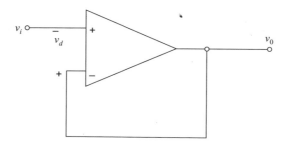

Fig. 5.7 Unity gain amplifier.

By writing loop equation from the circuit of the Fig. 5.7, we get:

$$v_i = v_0 - v_d = v_0 \left(1 - \frac{1}{A}\right)$$

or $\quad v_i = v_0$ as A is very high

$\therefore \quad v_0 = v_i$ ...(5.5)

5.8 OP-AMP AS ADDER OR SUMMER

An **adder** circuit is shown in Fig. 5.8. The three inputs v_1, v_2 and v_3 have series resistors R_1, R_2 and R_3 respectively. The currents are given by:

$$i_1 = \frac{v_1 - v}{R_1}, \quad i_2 = \frac{v_2 - v}{R_2} \quad \text{and} \quad i_3 = \frac{v_3 - v}{R_3}$$

$\therefore \quad i_f = \frac{v_1 - v}{R_1} + \frac{v_2 - v}{R_2} + \frac{v_3 - v}{R_3}$

and if

$$= \frac{v_0 - v}{R_f} \quad \text{as} \quad i \cong 0$$

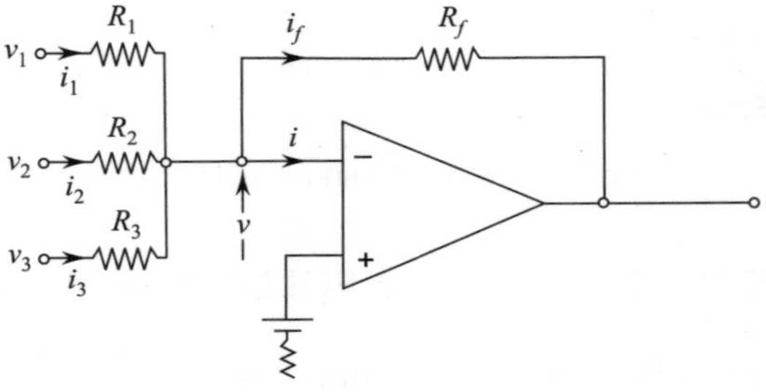

Fig. 5.8 Op-amp as an adder.

Normally, v is very small as compared to other voltages, hence, we get:

$$i_1 = \frac{v_1}{R_1}, \quad i_2 = \frac{v_2}{R_2} \quad \text{and} \quad i_3 = \frac{v_3}{R_3}$$

We know that:

$$-i_f = i_1 + i_2 + i_3 \quad \text{as} \quad i \cong 0$$

$\therefore \quad -\dfrac{v_0}{R_f} = \dfrac{v_1}{R_1} + \dfrac{v_2}{R_2} + \dfrac{v_3}{R_3}$

If $\quad R_f = R_1 = R_2 = R_3$, then we get:

$-v_0 = v_1 + v_2 + v_3$

or $\quad v_0 = -(v_1 + v_2 + v_3)$ ...(5.6)

Thus, apart from phase reversal, the output voltage is the sum of input voltages.

5.9 OP-AMP AS DIFFERENCE AMPLIFIER

Op-amp usage as **differential amplifier** is quite common. Differential amplifier amplifies the difference between two signals as Op-amp is a linear amplifier, the output is proportional to the difference in signal between two input terminals. Figure 5.9 shows differential amplifier circuit.

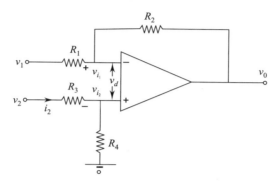

Fig. 5.9 Differential amplifier circuit.

If $v_2 = 0$, then the circuit becomes as shown in Fig. 5.10. Circuit becomes inverting type.

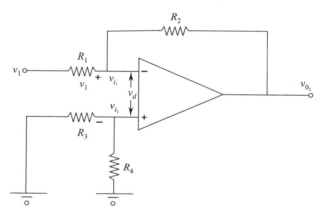

Fig. 5.10 Inverting Amplifier with $v_2 = 0$.

In this case, voltage output is:

$$v_{0_1} = -\frac{R_2}{R_1} v_1$$

Now, $v_1 = 0$, the +ve circuit becomes as shown in Fig. 5.11.

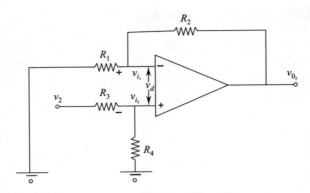

Fig. 5.11 Non-inverting amplifier with $v_1 = 0$.

In this case, circuit becomes inverting type, hence, we get:

$$\frac{v_{0_2}}{v_{i_2}} = A = \left(1 + \frac{R_2}{R_1}\right)$$

or

$$v_{0_2} = v_{i_2}\left(1 + \frac{R_2}{R_1}\right)$$

we also know that:

$$v_{i_2} = \left(\frac{R_4}{R_3 + R_4}\right)v_2$$

∴

$$v_{0_2} = \left(\frac{R_4}{R_3 + R_4}\right) \times v_2 \times \left(1 + \frac{R_2}{R_1}\right)$$

The value of full output is given by:

$$v_0 = v_{0_1} + v_{0_2}$$

or

$$v_0 = \Delta\left(\frac{R_2}{R_1}\right)v_1 + \left(\frac{R_4}{R_3 + R_4}\right) \times v_2 \times \left(1 + \frac{R_2}{R_1}\right)$$

$$= -\frac{R_2}{R_1} \times v_1 + \frac{\left(\frac{R_4}{R_3}\right)}{1 + \left(\frac{R_4}{R_3}\right)} \times v_2 \times \left(1 + \frac{R_2}{R_1}\right)$$

In case, $\dfrac{R_2}{R_1} = \dfrac{R_4}{R_3}$, we get:

$$v_0 = -\frac{R_2}{R_1} \times v_1 + \frac{\frac{R_2}{R_1}}{1 + \left(\frac{R_2}{R_1}\right)} \times v_2 \times \left(1 + \frac{R_2}{R_1}\right)$$

Operational Amplifier (Op-Amp)

or
$$v_0 = -\frac{R_2}{R_1} \times v_1 + \frac{R_2}{R_1} \times v_2$$

or
$$v_0 = \frac{R_2}{R_1}(v_2 - v_1)$$

Thus, a differential amplifier amplifies the difference between two voltages.

5.10 SUBTRACTOR

It **subtracts** one voltage from other. Consider Fig. 5.12 wherein $R_1 = R_2 = R_3 = R_4 = R$

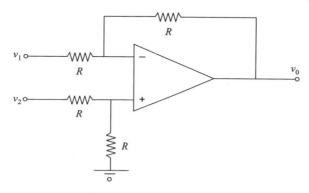

Fig. 5.12 Subtractor circuit.

Using **differential** amplifier principle, we get:

$$v_0 = \frac{R_2}{R_1}(v_2 - v_1)$$

or
$$v_0 = v_2 - v_1 \quad \text{as} \quad R_1 = R_2 = R$$

5.11 DIFFERENTIATOR

A differentiating circuit is shown in Fig. 5.13.

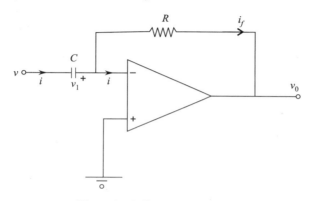

Fig. 5.13 Differentiator circuit.

From the circuit of the figure, we know:

$$i_1 = c\frac{dv}{dt}$$

But $i_{in} = 0$ for an Op-amp, hence, weget:

$$v_0 = -i_f R = -i_s R = -RC\frac{dv}{dt}$$

or
$$v_0 = -RC\frac{dv}{dt}$$

i.e., output Voltage is differentiation of input voltage.

5.12 INTEGRATOR

An **integrator circuit** is shown in Fig. 5.14.

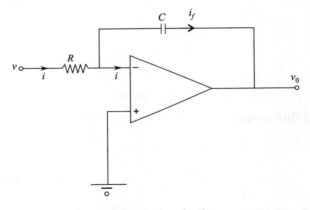

Fig. 5.14 Integrator circuit.

The current $\quad i = \dfrac{v}{R}$ since $i_i = 0$ V

$\therefore \quad v_0 = -\dfrac{1}{c}\int i_f \, dt = -\dfrac{1}{c}\int i \, dt$

By substituting $i = \dfrac{v}{R}$, we get:

$$v_0 = -\frac{1}{R_C}\int V_{dt}$$

Thus, the output is integral of the input.

5.13 OP-AMP PARAMETERS

5.13.1 Input Offset Voltage

Ideal amplifier gives zero output against zero input. In practice, there has to be a little voltage present at input, to get output as zero.

Figure 5.15 shows an amplifier wherein input voltage has some non-zero value to get zero output.

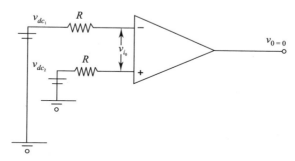

Fig. 5.15 Demonstration of input offset.

Thus,
$$v_{i_0} = v_{dc_1} - v_{dc_2}$$

5.13.2 Input Offset Current

The algebric sum of difference between two bias current inputs must be zero. Demonstration circuit for input offset current in shown in Fig. 5.16

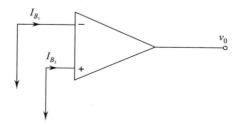

Fig. 5.16 Demonstration of input offset current.

The offset current,
$$I_{i_0} = \left| I_{B_1} - I_{B_2} \right|$$

It is used as an indicater of mismatching between two current.

5.13.3 Bias Current

Figure 5.16 shows demonstration of bias currents as well. Bias current average of current that flows between two treminals, hence,

bias current,
$$I_B = \frac{I_{B_1} + I_{B_2}}{2}$$

5.13.4 Slew Rate

It is the maximum rate of change of output voltage with respect to time. It is given in V/μs.

Slew rate, $$SR = \left.\frac{dv_o}{dt}\right|_{max} = V_p \text{ V/μs}$$

$$\therefore \left.\frac{dv_o}{dt}\right|_{max} = \frac{2\pi f V_p}{10^6} \text{ V/μs}$$

Where f = input frequency

V_p = peak value of output voltage.

5.13.5 Common Mode Rejection Ratio (CMRR)

This is a figure of merit for a differential amplifier and it is defined as:

$$CMRR = \frac{\text{differential voltage gain}}{\text{common mode voltage gain}}$$

or

$$CMRR = \frac{A_d}{A_{cm}}$$

where A_d = differential voltage gain

A_{cm} = common mode voltage gain.

A common mode circuit is shown in Fig. 5.17

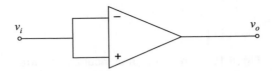

Fig. 5.17 Common-mode circuit.

Common voltage in applied to both input terminals,

In practice, A_d is very large and A_{cm} is very small, hence, CMRR is very large value.

SOLVED EXAMPLES

Example 5.1 For a given op-amp, CMRR = 105 and differential gain $A_d = 105$, Determine the common mode gain A_{cm} of the Op-amp.

Solution: Since CMRR is defined as

$$\text{CMRR} = \frac{A_d}{A_{cm}}$$

Common mode rejection ratio = $\dfrac{\text{Differential gain}}{\text{Common mode gain}}$

given that   CMRR = 105
and $A_d = 105$

therefore $A_{cm} = \dfrac{A_d}{\text{CMRR}} = \dfrac{105}{105} = 1$

Example 5.2 The output voltage of a certain op-amp ckt changes by 20 V in 4 micro sec. What is slew rate?

Solution: The slew rate,

$$SR = \frac{dv_0}{dt} = \frac{20 \text{ V}}{4 \text{ }\mu\text{sec.}} = 5 \text{ V}/\mu\text{sec.}$$

Example 5.3 Design a non-inverting amplifier ckt that is capable of providing a voltage gain of 10. Assume an ideal operational amplifies. (Resistor should not exceed 30 kΩ)

Solution:

Gain *AF* is given as

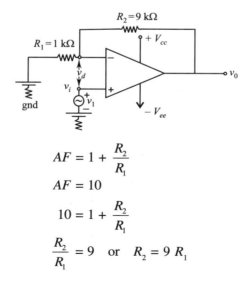

$$AF = 1 + \frac{R_2}{R_1}$$

given that $AF = 10$

$$10 = 1 + \frac{R_2}{R_1}$$

$$\frac{R_2}{R_1} = 9 \quad \text{or} \quad R_2 = 9 R_1$$

Taking
so that
$$R_1 = 1 \text{ k}\Omega$$
$$R_2 = 9 \text{ k}\Omega$$

$$\therefore \quad AF = 1 + \frac{R_2}{R_1}$$

$$AF = 1 + \frac{9}{1} = 10$$

$$AF = 10$$

Figure shows the required non-inverting amp.

Example 5.4 A 5 mV, 1 KHz Sinusoidal voltage is applied at the input of an Op-amp integrator for which $R = 100 \text{ k}\Omega$ and $C = 1 \text{ }\mu f$. Calculate the output voltage.

Solution: Given that
$$v_i = 5 \sin \omega t \text{ mV}$$
$$= 5 \sin 2\pi\omega t \text{ mV}$$
$$= 5 \sin 2000 \pi t \text{ mV}$$

We know that the output of an integrator is

$$v_o = \frac{-1}{RC} \int_0^t v_i \, dt = \frac{1}{100 \times 10^3 \times 1 \times 10^{-6}} \int_0^t 5 \sin 2000 \pi \, t dt$$

$$v_o = \frac{-1}{0.1} \int_0^t 5 \sin 2000 \pi \, t dt = -10 \left[\frac{-5 \cos 2000 \pi t}{2000 \pi} \right]_0^t$$

$$v_o = -50 \left[\frac{\cos 2000 \pi t - 1}{2000 \pi} \right] = \frac{1}{40 \pi} (\cos 2000 \pi t - 1) \text{ mV}.$$

Example 5.5 Show that the output of the inverting integrator of given figure is the time integral of the Input signal, assuming the op-amp is ideal.

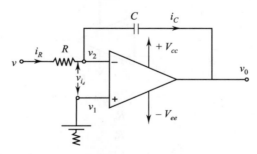

Solution: We know that for an Op-amp

$$A = \frac{v_0}{v_{id}}$$

Operational Amplifier (Op-Amp)

Assuming, op-amp is an ideal one:

$$A \cong \infty, \quad v_{id} = \frac{v_0}{A} = \frac{v_0}{\infty} = 0$$

i.e. $\quad v_{id} = 0$

i.e. $\quad v_1 - v_2 = 0 \quad \text{or} \quad v_1 = v_2$

Op-amp is ideal, it will draw zero current.

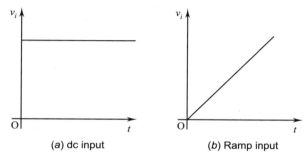

(a) dc input (b) Ramp input

Therefore, $\quad i_R = i_C$ from figure the value of

$$i_R \text{ as } \frac{v_1 - v_2}{R} = i_C \quad \text{But} \quad i_C = 0$$

$$\therefore \quad \frac{v_i - 0}{R} = i_C \quad \text{or} \quad i_C = \frac{v_i}{R}$$

But i_C = current through capacitor

We know that the current through a capacitor may be expressed as

$$i_C = C \cdot \frac{dv_c}{dt}$$

Therefore, $\quad C \cdot \frac{dv_c}{dt} = \frac{v_i}{R}$ (here v_c is the V across C)

given as $\quad v_c = v_2 - v_0$

Substituting the value of v_c, we get

$$C \cdot \frac{d}{dt}[v_2 - v_0] = \frac{v_i}{R} \quad \text{But } v_2 = 0$$

Hence $\quad C \cdot \frac{d}{dt}[0 - V_0] = \frac{v_i}{R}$

or $\quad \frac{v_i}{R} = C \cdot \frac{d}{dt}(-v_0)$

Integrating both sides of above equations, w.r.t. we get

$$\int_0^t \frac{v_i}{R} dt = C \cdot (-v_0) + A$$

$$v_0 = \frac{-1}{RC} \int_0^t v_i \, dt + A$$

$$v_0 = \frac{-1}{RC} \int_0^t v_i \, dt + A \qquad \ldots(i)$$

where A is the integration constant and is proportional to the value of the output voltage v_0 at time $t = 0$ sec. From *eqn.* (*i*), it is clear that the output voltage v_0 is equal to the integration of input voltage v_i.

Example 5.6 In the given figure the variable resistance varies from zero to 100 kW. Find out the maximum and the minimum closed loop voltage given.

Solution: The given ckt is a non-inverting Op-amp amplifier

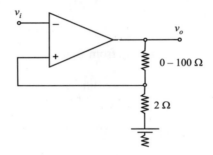

Therefore,

$$AF = \frac{v_0}{v_i} = 1 + \frac{R_F}{R_1}$$

or

$$v_0 = v_i \left[1 + \frac{R_F}{R_1} \right]$$

But for $\qquad R_F = 0, R_1 = 2 \text{ k}\Omega$

Hence using (i), we get

$$v_0 = v_i \left[1 + \frac{0}{R_1} \right]$$

$$v_0 = v_i (1 + 0) = v_i$$

Then minimum closed loop voltage gain

$$AF_{min} = \frac{v_0}{v_i} = 1$$

Similarty for $\qquad R_F = 100 \text{ k}\Omega, R_1 = 2 \text{ k}\Omega$

Therefore, using (i), we get

$$v_o = v_i\left[1 + \frac{100}{2}\right] = v_i \cdot 51$$

$$v_o = 51\, v_i \quad \text{or} \quad \frac{v_o}{v_i} = 51$$

Then maximum closed loop voltage gain

$$AF_{max} = \frac{v_o}{v_i} = 51.$$

Example 5.7 Find an expression for the output v_0 of the amplifier ckt of given figure. Assume Op-amp is ideal. What mathematical operation does this ckt perform?

Solution: Making $v_B = 0$, we have

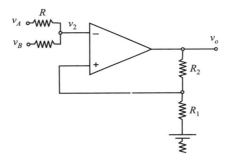

$$V_2 = \frac{v_R \cdot R}{R + R} = \frac{v_A}{2}$$

Let with $v_B = 0$, the value of v_0 will be v_{0_1}

then

$$v_{0_1} = \left(1 + \frac{R_2}{R_1}\right) v_2$$

$$v_{0_1} = \left(1 + \frac{R_2}{R_1}\right) \frac{v_A}{2} \quad [\because \text{It is a non-inverting amplifier}]$$

or

$$v_{0_1} = \left(1 + \frac{R_2}{R_1}\right) \frac{v_A}{2} \qquad \ldots(i)$$

Similarly, making $v_A = 0$, we have

$$v_{0_2} = \left[1 + \frac{R_2}{R_1}\right] \frac{v_B}{2} \qquad \ldots(ii)$$

Applying principle of superposition, the output voltage v_0 of amplifier will be

$$v_o = v_{0_1} + v_{0_2}$$

$$= \left(1 + \frac{R_2}{R_1}\right)\frac{v_A}{2} + \left[1 + \frac{R_2}{R_1}\right]\frac{v_B}{2}$$

or

$$v_o = \frac{1}{2}\left[1 + \frac{R_2}{R_1}\right](v_A + v_B)$$

From above equation, it is clear that the given ckt performs the mathematical operation of a non-inverting adder.

Example 5.8 For difference mode, gain of an amplifier is $A = 2000$ and CMRR = 10,000. Calculate output voltage V out when $v_1 = 1.0$ mV and $v_2 = 0.9$ mV.

Solution: Given that

$$v_1 = 1.0 \text{ mV}, v_2 = 0.9 \text{ mV}$$

Difference val
$$v_d = v_1 - v_2$$
$$v_d = 1.0 - 0.9 = 0.1 \text{ mV}$$

Also, common-mode voltage will be

$$v_c = \frac{v_1 + v_2}{2} = \frac{1.0 + 0.9}{2} = 0.95 \text{ mV}$$

Difference Voltage gain $\quad A = 2000$

and $\quad$ CMMR = 10,000

The output voltage is expressed as

$$v_o = A\, v_d \left[1 + \frac{1}{CMRR}\frac{v_C}{v_d}\right]$$

Substituting all the values, we get

We get
$$v_o = 2000 \times 0.1 \text{ mV} \left[1 + \frac{1}{10.000} \times \frac{0.95}{0.1}\right]$$

$$v_o = 200.19 \text{ mV}.$$

Example 5.9 Figure shows an Op-amp. Obtain the value o/p voltage in steady-state condition where (i) switch S is open and (ii) switch S is closed.

Solution: Case (i) when switch S is open then $R_F = 1 + 1 = 2$ kΩ

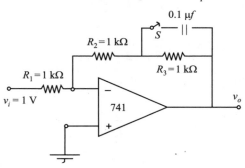

Therefore, output voltage

$$v_o = -v_i \times \frac{R_F}{R_1}$$

$$= -1 \times \frac{2\ k\Omega}{1\ k\Omega}$$

$$v_o = -2 \text{ volt}$$

Case (ii) when the switch S is closed then the impedance at the feedback ckt. will be

$$Z_F = \frac{R_3}{1+j\omega c.R_3} + R_2 = \frac{R_3 + R_2(1+j\omega c.R_3)}{1+j\omega cR_3}$$

Therefore

gain
$$A_F = \frac{-Z_F}{R_1} = -\frac{R_3 + R_2(1+j\omega cR_3)}{(1+j\omega cR_3)R_1}$$

Now, since the gain at the steady-state corresponds to d.c. or low frequency, therefore neglecting $j\omega$ terms, we get

gain
$$AF = -\frac{R_3 + R_2}{R_1} = -\frac{1\ k\Omega + 1\ k\Omega}{1\ k\Omega} = -2$$

Hence the output voltage $v_o = A_F \cdot v_i = -2$ V.

Exapmple 5.10 An Op-amp has feed back resistor $R_F = 12$ kΩ and the resistor in the Input sides are $R_1 = 12$ kΩ, $R_2 = 2$ kΩ and $R_3 = 3$ kΩ. The corresponding inputs are $v_{t_1} = +9$ V, $v_{t_2} = -3$ V and $v_{t_3} = -1$ V.

Non-Inverting terminal is grounded. Calculate the output voltage.

Solution: Given

$$R_F = 12\ k\Omega,\ R_1 = 12\ k\Omega,\ R_2 = 2\ k\Omega \text{ and } R_3 = 3\ k\Omega$$
$$v_1 = 9 \text{ V};\ v_2 = -3 \text{ V and } v_3 = -1 \text{ V}$$

Output voltage

$$v_o = -R_F \left[\frac{v_1}{R_1} + \frac{v_2}{R_2} + \frac{v_3}{R_3}\right]$$

$$= -12\ k\Omega \left[\frac{9}{12\ k\Omega} + \frac{-2}{2\ k\Omega} + \frac{-1}{3\ k\Omega}\right]$$

$$v_o = -9 + 12 + 4 = 7 \text{ V}.$$

Example 5.11 Realise a ckt – to obtain $v_0 = -2v_1 + 3v_2 + 4v_3$ using operational amplifier. Use minimum value of resistance as 10 kΩ.

Solution: For an operational amplifier we have

$$v_o = -\left[\frac{R_F}{R_1}v_1 + \frac{R_F}{R_2}v_2 + \frac{R_F}{R_3}v_3\right]$$

Comparing the above expression with the given expression for the output

i.e.
$$v_o = -2v_1 + 3v_2 + 4v_3$$
$$= -[2v_1 - 3v_2 - 4v_3]$$

We have, $\dfrac{R_F}{R_1} = 2$; $\dfrac{R_F}{R_2} = 3$ and $\dfrac{R_F}{R_3} = 4$

Resistance R_3 will be of minimum value of 10 kΩ.

Thus
$$R_F = 4R_3 = 4 \times 10 = 40 \text{ k}\Omega$$

$$R_2 = \dfrac{R_F}{3} = \dfrac{40}{3} = 13.33 \text{ k}\Omega$$

$$R_1 = \dfrac{R_F}{2} = \dfrac{40}{2} = 20 \text{ k}\Omega.$$

Example 5.12 Given figure shows a non-inverting op-amp summer with $v_1 = 2$ V and $v_2 = -1$ V. Calculate the output voltage v_0.

Solution: According to superposition theorem, we have

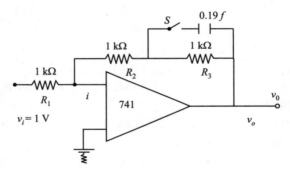

$$v_o = v_o' + v_o''$$

when v_o' is the output produced by v_1 of + 2 V and v_o'' is the output produced by
$$v_2 = -1 \text{ V}$$

When $v_2 = -1$ V is made zero, Input at Non-inverting I/p terminal will be

$$v_{s_1} = v_1 \times \dfrac{R \| R}{R + (R \| R)}$$

$$= 2 \times \dfrac{R/2}{R + R/2} = 2/3 \text{ V}$$

$$v_o' = v_{s_1} \left[1 + \dfrac{R_F}{R}\right] = \dfrac{2}{3}\left[1 + \dfrac{2R}{R}\right] = 2 \text{ V}$$

When $v_1 = 2$ V is made zero, we have

$$v_{s_2} = v_2 \times \frac{R_1 \parallel R}{R + (R_1 \parallel R)}$$

$$= -1 \times \frac{R/2}{R + R/2} = \frac{-1}{3} \text{ V}$$

and

$$v_0'' = v_{s_2}\left[1 + \frac{R_F}{R}\right]$$

$$= \frac{-1}{3}\left[1 + \frac{2R}{R}\right] = -1 \text{ V}$$

Hence O/P voltage $v_o = v_0' + v_0''$
$= 2 + (-1) = 1$ V.

Example 5.13 If the non-ideal Op-amp of the ckt. of following figure has an open loop gain.

$$A_{OL} = -10^4. \text{ Find } v_0$$

Solution: It is an inverting Op-amp because Input is applied at the inverting terminal while other terminal is grounded.

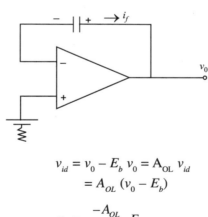

$$v_{id} = v_0 - E_b, v_0 = A_{OL} v_{id}$$
$$= A_{OL} (v_0 - E_b)$$

$$v_0 = \frac{-A_{OL}}{1 - A_{OL}} E_b$$

$$= \frac{-10^4}{1 - (-10^4)}$$

$[v_0 = 0.99 \, E_b]$.

Example 5.14 The output voltage of the summer is shown below. Calculate the value of feedback resistance.

Solution: The output of a summer ckt. is given by:

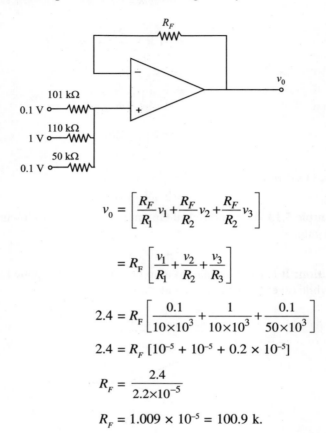

$$v_0 = \left[\frac{R_F}{R_1} v_1 + \frac{R_F}{R_2} v_2 + \frac{R_F}{R_2} v_3 \right]$$

$$= R_F \left[\frac{v_1}{R_1} + \frac{v_2}{R_2} + \frac{v_3}{R_3} \right]$$

$$2.4 = R_F \left[\frac{0.1}{10 \times 10^3} + \frac{1}{10 \times 10^3} + \frac{0.1}{50 \times 10^3} \right]$$

$$2.4 = R_F [10^{-5} + 10^{-5} + 0.2 \times 10^{-5}]$$

$$R_F = \frac{2.4}{2.2 \times 10^{-5}}$$

$$R_F = 1.009 \times 10^{-5} = 100.9 \text{ k}.$$

Example 5.17 Design a ckt. to give a weighted average $\frac{x}{3} + \frac{y}{2} + \frac{z}{6}$ where x, y and z are input voltage. What input resistors do you need if the feedback resister is 60 k ?

Solution: Let $v_1 = x$, $v_2 = y$, $v_3 = z$, then the output is

$$v_0 = -\left[\frac{R_f}{R_1} v_1 + \frac{R_f}{R_2} v_2 + \frac{R_f}{R_3} v_3 \right]$$

$$= -\left[\frac{60 \text{ k}\Omega}{R_1} x + \frac{60 \text{ k}\Omega}{R_2} y + \frac{60 \text{ k}\Omega}{R_3} z \right]$$

$$= -\left[\frac{x}{3} + \frac{y}{2} + \frac{z}{6} \right]$$

$$\therefore \qquad \frac{60\,k\Omega}{R_1} x = \frac{x}{3}, \frac{60\,k\Omega}{R_2} y = \frac{y}{2}, \frac{60\,k\Omega}{R_3} z = \frac{z}{6}$$

$$\Rightarrow \qquad R_1 = 180\ \Omega,\ R_2 = 120\ \Omega,\ R_3 = 360\ \Omega$$

Here input resistances are need if $R_1 = 180\ \Omega$, $R_2 = 100\ \Omega$ and $R_3 = 36\ \Omega$, then feed back resistance Rs.60 Ω.

Example 5.18 For an inverting amplifier in the given figure $R_1 = 1\ \Omega$ and $R_f = 100\ k\Omega$. Assuming an ideal amplifier, determine (*i*) the voltage gain, (*ii*) input impedence and (*iii*) the output impedence.

Solution: For the ideal inverting Op–amp we have:

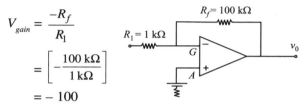

$$V_{gain} = \frac{-R_f}{R_1}$$

$$= \left[-\frac{100\,k\Omega}{1\,k\Omega} \right]$$

$$= -100$$

Since, point A in ground potential virtually, therefore, the impedence $z_{in} = R_1 = 1\ k\Omega$ and the output impedence of the ckt. equals the output impedence of the operaional amplifier.

Here output impedence of Op-amp is zero.

$$\therefore \qquad 2out = 0.$$

Example 5.19 A non-inverting amplifier in given figure is to be applied with a gain of 1.5 of $R_1 = 4\ k\Omega$, what value of R_f should be used?

Solution: We know that

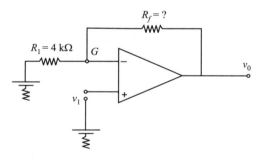

V_{gain} of the non-inverting amplifier

$$\frac{v_0}{v_1} = 1 + \frac{R_f}{R_1}$$

$$\left(\frac{v_0}{v_1} - 1 \right) = \frac{R_f}{R_1}$$

∴ $$R_f = R_1\left(\frac{v_0}{v_1} - 1\right)$$
$$= 4 \text{ k}\Omega \ (1.5-1)$$
$$= 4 \text{ k}\Omega \ (0.5) = 2.0 \text{ k}\Omega.$$

Example 5.20 Find the output voltage for the inverting summing ckt. of given below for $R_1 = 5$ kΩ, $R_2 = 3$ kΩ, $R_f = 5$ kΩ, $V_1 = 5 \sin \omega t$, $V_2 = 6 \sin \omega t$ and $U_3 = -5 \sin \omega t$.

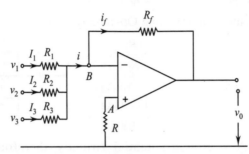

Solution: The output voltage of an inverting summing ckt. with three I/P is given by:

$$v_0 = \left[\frac{R_f}{R_1}v_1 + \frac{R_f}{R_3}v_2 + \frac{R_f}{R_2}v_3\right]$$

Using the value given then we get

$$v_0 = \left(\frac{-5}{5} \times \sin \omega t + \frac{6}{3} \times 6 \sin \omega t + \frac{5}{2} \times (-5 \sin \omega t)\right)$$
$$= -[+5 + 10 - 12.5]\sin \omega t$$
$$= -2.5 \sin \omega t.$$

Example 5.21 A subtracting amp or difference ckt. of figure has $v_1 = 60 \cos \omega t$, $v_2 = 18 \cos \omega t$ volt, $R_1 = R_f = 5$ kΩ and $R_2 = R_3 = 10$ kΩ. Find the value of v_0.

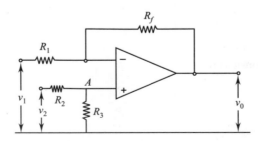

Solution: From given figure $$\frac{v_1 - v_2 R_2/R_2 + R_3}{R_1} = \frac{\frac{v_2 v_3}{(R_2 + R_3)} - v_0}{R_f}$$

for $R_1 = R_f$ and $R_2 = R_3$, we get

$$\frac{v_1 - v_2/2}{R_f} = \frac{v_2/2 - v_0}{R_f}$$

or
$$v_0 = v_2 - v_1 = 18 \cos \omega t - 6 \cos \omega t = 12 \cos \omega t.$$

Example 5.22 For an integrated ckt. of given figure $v_1 = (1 \text{ V}) \sin \omega t$. If $R_1 = 5 \text{ k}\Omega$ and $C = 1.0 \times 10^3$ PF. Find v_0 at $\omega t = \frac{\pi}{2}$, if $v_0(0) = 0$ and $\omega = 1$ MHz.

Solution: The o/p potential of the integrator ckt.

$$v_0 = \frac{1}{R_1 C} \int_0^t v_1 dt \text{ given } v_0(0) = 0$$

In the present exercises.

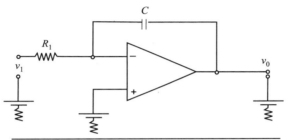

$R_1 = 5 \text{ k}\Omega, C = 1 \times 10^3$ PF

$v_1 = (1 \text{ V}) \sin \omega t$ and $\omega t = \frac{\pi}{2}$ or $t = \frac{\pi}{2\omega}$

$\therefore$
$$v_0 = \frac{-1}{R_1 C} \int_0^t (1 \text{ V}) \sin \omega t \, dt$$

$$= \frac{1}{R_1 C \omega} [\cos \omega t]_0^{\pi/2\omega}$$

$$= \frac{1}{R_1 C \omega} \left[\cos \frac{\pi}{2} - \cos 0 \right] = \frac{-1}{R_1 C \omega}$$

$$= -\frac{1}{5 \times 10^3 \times 1 \times 10^{-5} \times 10^6} = -0.2 \text{ V}$$

Example 5.23 In the given figure shown an active differentiator which has a very high input impedence. The second stage is an inverting buffer. If the given o/p signal is a triangular wave with its slope of ± 400 mV/20 µ sec. Find the output voltage.

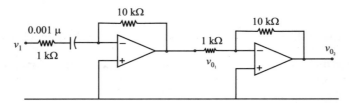

Solution: The input of a triangular wave, therefore the output must be a square wave. The output of the differentiator ckt. is given by the relation.

Here, $C = 0.001$ µF, $R_F = 10$ kΩ, $\dfrac{dv_I}{dt} = \pm \dfrac{400 \text{ mV}}{20 \text{ µsec}}$

$$v_{0_1} = -10 \times 10^3 \times 10^{-9} (\pm 400 \times 10^{-3} \times 20 \times 10^{-6})$$
$$= \pm 200 \text{ mV}$$

It will acts as o/p for the second stage, which is an inverting buffer.

$$= \text{the output voltage } v_{0_2} = -(R_2/R_1)v_i$$
$$= -10 (\pm 200 \text{ mV})$$
$$= \pm 2 \text{ V}.$$

Example 5.24 It is desired to have an output which is sum of the integrals of the various Input, i.e.;

$$e_0 = \int e_1\, dt + \int e_2\, dt + \int e_3\, dt + \ldots$$

Give an appropriate ckt and prove the result.

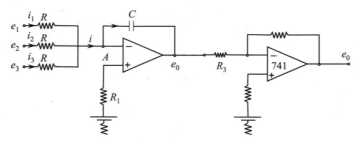

Solution: From the given ckt.

$$I = I_1 + I_2 + I_3 \ldots = -C\dfrac{de_0}{dt}$$

$$= \dfrac{e_1}{R} + \dfrac{e_2}{R} + \dfrac{e_3}{R} = -C\dfrac{de_0}{dt}$$

or

$$e_{0_1} = \dfrac{1}{R_c}\left[\int e_1 dt + \int e_2 dt + \int e_3 dt + \ldots\right]$$

Selecting $\quad R = 10\ k\Omega$ and $C = 100\ \mu F$ we get $RC = 1$

$$e_{o_1} = [\int e_1 dt + \int e_2 dt + \int e_3 dt ...]$$

if $\quad \dfrac{R_4}{R_3} = 1$ then $e_0 = -e_{0_1} = \int e_1\,dt + \int e_2\,dt + \int e_3\,dt$

and hence, the result.

Example 5.25 Design an Op-amp based non-inverting op-amp heavy a gain of 11. Determine the input impedence of this it the chosen op-amp has open loop gain of 10,000 and open loop input impedence of 1 $\mu\Omega$.

Solution: Figure shows the basic non-inverting amplifier using the op-amplifier. The gain of this amplifier is given by

$\left(1 + \dfrac{R_2}{R_1}\right)$ required gain = 11

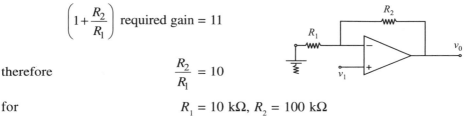

therefore $\quad \dfrac{R_2}{R_1} = 10$

for $\quad R_1 = 10\ k\Omega,\ R_2 = 100\ k\Omega$

Now for the non inverting amplifier the inpedance R_m is given by

open loop input impedance × loop gain = open loop impedance × $\dfrac{\text{open loop gain}}{\text{closed loop gain}}$

$$= 1 \times \dfrac{10000}{11}\ \mu\Omega$$

$$= 9091\ \mu\Omega.$$

Example 5.26 Design an Op-amp based logging type phase shifter that can shift the phase of an input sinosoidal signal by – 60 with a gain of unity. If the input signal has a peak amplitude of 5 volts and the highest input frequency 50 kHz, what should be the slew rate of the choosen op-amp so that it does not limit the bandwidth?

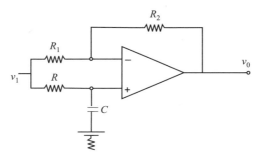

Solution: For unity gain $R_1 = R_2$

The Phase Shift introduced by the ckt (θ) is given by

$$\theta = -2 \tan^{-1} (2\pi f RC)$$
$$60° = -2 \tan^{-1} (2\pi \times 50 \times 10^3 RG)$$
$$10^5 \pi RC = \tan 30 = 0.577$$

for

$$RC = \frac{0.577}{10^5 \pi} = 0.184 \times 10^{-5}$$

$$C = 0.001 \ \mu F$$

if

$$R = \frac{0.184 \times 10^{-5}}{0.001 \times 10^{-6}} = 1840° = 1.84 \ kW$$

Let $R_1 = R_2 = 10 \ k\Omega$

I/P signal amp = 5 V peak

$$f_{max} = \frac{\text{slew rate}}{2\pi v_{o(max)}}$$

or slew rate

$$= 2\pi \times 5 \times 50 \times 10^3 = 31.4 \times 50 \times 10^3$$
$$= 157 \times 10^4 \ v/s = 1.57 \ \mu \ sec.$$

Example 5.27 For the non-inverting ckt. $R_L = 5 \ k\Omega$ and $R_f = 200 \ k\Omega$ determine the v_{gain}.

Solution: (Ar) $v_{gain} = 1 + \dfrac{R_F}{R_L} = 1 + \dfrac{200}{5} = 41.$

Example 5.28 Find the output of the given circuit.

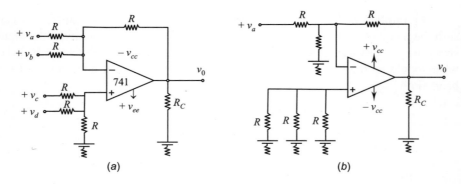

(a)   (b)

Solution: The o/p voltage equation for this ckt can be obtained by using the superposition theorem. For instance to find the o/p voltage due to v_a along, reduce all the I/P voltage v_b, v_c, v_d to zero v_d.

Infact this ckt. is an inverting amplifier in which the inverting input is at virtual ground ($v_2 = 0$ V) So, $v_{0_a} = - R v_a = - v_a$. Now it input voltage v_a, v_b and v_d

are set R to zero, the ckt. is become a inverting amplifier in which the voltage v_1 at the non-inverting input pin

$$v_1 = \frac{v_c \times R_{12} - v_{cb}}{R + R_{12}}$$

This means that the o/p voltage due to v_c alone is $v_{oc} = \left(1 + \frac{R}{R_1}\right) v_1 = v_c$

Similarly = $v_{od} = v_d$. Thus, by super position theorem the o/p voltage due to all form voltage is given by

$$v_0 = v_0 a + v_0 b + v_0 c + v_0 d$$
$$v_0 = -v_a - v_b + v_c + v_d$$

Example 5.29 If time constant of the integration is one sec and input is a step (*dc*) voltage as shown in the figure. Determine the o/p voltage and sketch it assume that the op-amp is initially null.

Solution: The function is constant beginning at two seconds.

This is $\qquad v_i = 2$ V for $0 \leq t \leq 4$ therefore.

$$t = 4$$

$$v_0 = -\int_0^4 2 dt$$

$$= -\left[\int_0^1 2dt + \int_1^2 2dt + \int_2^3 2dt + \int_3^4 2dt\, v_0\right]$$

$$= -(2 + 2 + 2 + 2) = -8 \text{ V}$$

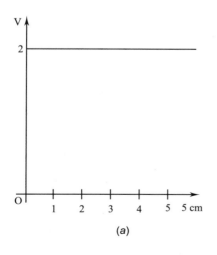

(a)

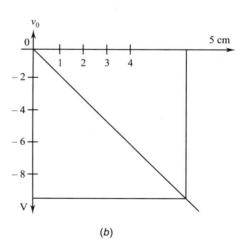

(b)

Example 5.30 Sketch all output waveform of the differential amplifier if input is

(i) sine wave

(ii) square wave

Solution:

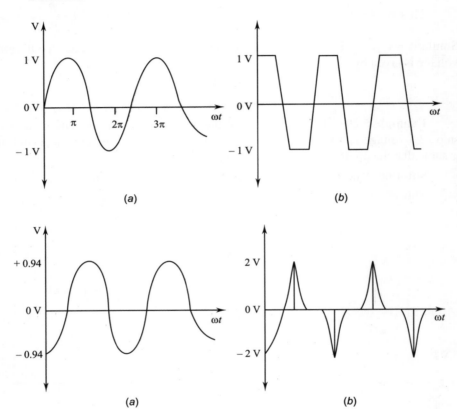

SUMMARY

1. **Basics:** An op-amp has inverting and non-inverting inputs. It has high imput inpedence at both input terminals and a low output impedence. It has large voltage gain which remains constant over a large frequency range.

2. **Inverting Amplifier**

$$A_v = -\frac{R_f}{R_1}$$

3. **Non-inverting Amplifier**

$$A_v = 1 + \frac{R_f}{R_1}$$

4. **Voltage follower** amplifier gives unity voltage gain.

5. **Adder**

$$v_0 = -(v_1 + v_2 + v_3)$$

6. **Difference Amplifier**

$$v_0 = \frac{R_2}{R_1}(v_2 - v_1)$$

7. **Subtractor**

$$v_0 = v_2 - v_1 \quad \text{for } R_1 = R_2$$

8. **Differentiator**

$$v_0 = -R_C \frac{dv}{dt}$$

9. **Integrator**

$$v_0 = \frac{1}{R_C}\int v\, dt$$

10. **Op-amp parameters.**

    (*i*) Input offset voltage

    $$v_{i_0} = v_{dc_1} - v_{dc_2}$$

    (*ii*) Input offset current

    $$I_{i_0} = |I_{B_1} - I_{B_2}|$$

    (*iii*) Bias Current

    $$I_B = \frac{I_{B_1} + I_{B_2}}{2}$$

    (*iv*) Slew Rate

    $$SR = \frac{2\pi f v_P}{10^6} v/\mu s$$

    (*v*) $CMMR = \dfrac{A_d}{A_{cm}}$

EXERCISES

5.1 Explain an ideal op-amp with circuit diagrams.

5.2 What do you understand by inverting and non-inverting in an op-amp? Explain in detail.

5.3 Derive equation for an adder.

5.4 What is output from a differentiator? Derive the formula.

5.5 Describe integrators and differentiators. Give their output equations.

5.6 What is an Op-amp? Draw and explain the block diagram of op-amp.

5.7 Explain the significance of virtual ground in a basic inverting amplifier. How would you explain its existence?

5.8 Draw the pin diagram of an IC used as an operational ampr. Explain the functions of the different pin connection.

5.9 Explain the functioning of a buffer ampr.

5.10 Explain why open loop Op-amp configuration are not used in linear applications.

5.11 Find the output voltage of an open looped Op-amp having $A = 200000$ when the differential I/P voltage is $\pm 50\ \mu V$. [**Ans.** ± 10 V]

5.12 For a given op-amp, CMRR $= 10^5$ and differential gain $A_d = 10^5$. Determine the common mode gain A_{cm} of the op-amp. [**Ans.** 1]

5.13 The output voltage of a certain op-amp ckt changes by 20V in 4 sec. What is its slew rate? [**Ans.** $5\ \mu$ sec]

5.14 In the ckt of fig. if $R_1 = R_2 = 1$ kΩ, $R_f = R_3 = 10$ kΩ, $v_d = 5$ mV sine wave at 1 kHz and V_m (noise voltage) $= 2$ mV at 60 Hz, calculate (a) the output voltage at 1 kHz and (b) the amplitude of the induced 60 Hz noise at the output. The Op-amp is the m741 with CMRR (dB) $= 90$dB.

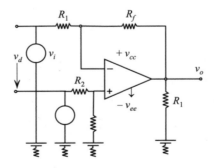

[**Ans.** 50 mV, 0.63 mV (at 60 Hz.)]

5.15 Find the closed loop ckt. gain and the output voltage for an inverting Op-amp having input signal v_i as 1 V (peak to peak) and R_1 and R_f as 1 kΩ and 10 kΩ respectively. [**Ans.** 10 V]

5.16 Find the closed loop ckt gain and the output voltage for a non-inverting Op-amp ckt having input signal V in as 1 V (peak to peak) and R_1 and R_f as 1 kΩ and 10 kΩ respectively. [**Ans.** 11, 11 V]

5.17 An operational amplifier is to have a voltage of 50. Calculate the required values for the external resistors R_1 and R_f if (a) a non-inverting (b) an inverting gain is required. [**Ans.** 2 kΩ, 2 kΩ]

5.18 In a differential amplifier of the type shown in fig. $R_1 = 10$ kΩ, $R_2 = 100$ kΩ, $R_3 = 10$ kΩ and $R_4 = 10$ kΩ. Calculate the output voltage of the ckt if (i) $v_1 = 10$ mV, $v_2 = 0$, (ii) $v_1 = 0$, $v_2 = 10$ mV (iii) $v_1 = 100$ mV, $v_2 = 50$ mV, and (iv) $v_1 = 50$ mV, $v_2 = 1$ mV. [**Ans.** −100 mV + 100 mV, −500 mV + 500 mV]

5.19 Find an expression for the Input impedance of the unity follower amplifier.

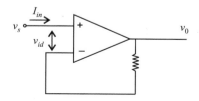

[**Ans.** $z_{in} = -AOL\, R_d$]

5.20 Find an expression for the output v_0 of the ampr ckt. of the given figure Assume an ideal Op-amp. What mathematical operation does the ckt. perform?

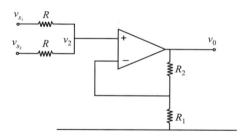

[**Ans.** $\dfrac{1}{2}\left[1+\dfrac{R_2}{R_1}\right](v_{s_1}+v_{s_2})$]

5.21 For the non-inverting amplifier of figure, find an exact expression for the v_{gain} ratio.

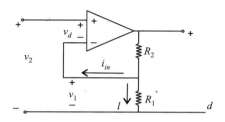

[**Ans.** $A_v = \dfrac{v_0}{v_2} = \dfrac{R_1+R_2}{1-\dfrac{R_1 R_2}{A_{OL}\cdot R_d}-\dfrac{(R_1+R_2)}{A_{OL}}}$]

5.22 Find the V gain ratio A_v of the non-inverting ampilifier of given figure in terms of its CMRR. Assume $v_1 = v_2$ in so far as the common mode gain is concerned.

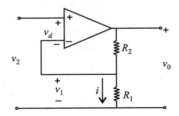

$$\left[\text{Ans. } A_v = \frac{v_0}{v_2} = \frac{-A_{OL}}{1 - \dfrac{A_{OL} R_1}{R_1 + R_2}} - \frac{A_{OL}}{\text{CMRR} \left(1 - \dfrac{A_{OL} R_1}{R_1 + R_2}\right)}\right]$$

5.23 An inverting summer (fig) has n input with $R_1 = R_2 = R$. Assume that the open loop basic Op-amp gain A_{OL} is infinite, but that the inverting terminal input current is negligible. Derive a relationship that shows how gain magnitude is reduced in the presence of multiple inputs for a practical Op-amp.

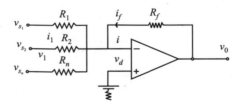

$$\left[\text{Ans. } A_n = -\frac{R_f / R}{1 - \dfrac{n R_f}{(1 + R) A_{OL}}}\right]$$

6

SWITCHING THEORY AND LOGIC DESIGN (STLD)

6.1 INTRODUCTION

Switching circuits are for the use of binary variables and application of binary logic. Electronic digital circuits are also types of switching circuits. In digital systems, the numbers are represented by binary numbers rather than decimal system. The binary numbers are also used in arithmetic operations. Digital circuits use binary signals to control conduction or non-conduction or non-conduction state of an active element such as transistor, FET, MOSFET, *etc*. Digital circuits use transistor as switch. Switching circuits are also called logic circuits as it can establish logical manipulation with proper controlled inputs. Logic circuits are used to compute and control any desired information in the form of binary signals. Logic circuits which perform logical functions are called logic gates. Logic gates are the basic building blocks of any combinational logic network as per requirement.

6.2 NUMBER SYSTEM

6.2.1 Decimal System

Decimal system is normally used for expressing numbers. In decimal system, there are ten digits from 0 to 9, therefore, it has a base of 10. This implies that each digit in a decimal number represents a multiple of a power of 10.

Consider decimal number 468. It has four hundreds, six tens and eight units. This can be written as:
$$4 \times 10^2 + 6 \times 10^1 + 8 \times 10^0$$

A weight is assigned to the postion of each digit. Whole numbers have weights which are positive, increasing from right to left. The lowest is $10^0 = 1$. In the case of fractional numbers the weights are negative decreasing from left to right, starting with $10^{-1} = 0.1$. Hence, the number 268.17 can be written as:

$$2 \times 10^2 + 6 \times 10^1 + 8 \times 10^0 + 1 \times 10^{-1} + 7 \times 10^{-2}$$

6.2.2 Binary System

Binary system has a base of 2 and there are only two digits, 0 and 1. These are called bits, thus a binary number can only be expressed in 0's and 1's. For example, binary number 101 can be written as:

$$1 \times 2^2 + 0 \times 2^1 + 1 \times 2^0$$

In decimal system, this represents $4 + 0 + 1 = 5$. Table 6.1 shows four digit binary numbers. It may be noted that number of bits has to be increased to represent larger numbers. 5 bits would double the range compared to 4 bits. 6 bits would double the range of 5 bits. It can be seen from the table that $1101_2 = 13_{10}$.

Table 6.1 Four-digit binary numbers

Decimal	Binary			
	$2^3 = 8$	$2^2 = 4$	$2^1 = 2$	$2^0 = 1$
0	0	0	0	0
1	0	0	0	1
2	0	0	1	0
3	0	0	1	1
4	0	1	0	0
5	0	1	0	1
6	0	1	1	0
7	0	1	1	1
8	1	0	0	0
9	1	0	0	1
10	1	0	1	0
11	1	0	1	1
12	1	1	0	0
13	1	1	0	1
14	1	1	1	0
15	1	1	1	1

A binary number is also a weighted number similar to the decimal system. The least significant bit (LSB) is the right hand bit and it has a weight $2^0 = 1$. Increase of weights are from right to left by a power of 2 for each bit. The most significant bit (MSB) is the left hand bit. The size of the binary number determines the weight.

Representation of fractional binary numbers are done by placing bits to the right of the binary number, hence, Table 6.2 gives details.

Table 6.2 Binary fractional number bits

Binary	2^6	2^5	2^4	2^3	2^2	2^1	2^0	2^{-1}	2^{-2}	2^{-3}	2^{-4}
Decimal	64	32	16	8	4	2	1	0.5	0.25	0.125	0.625

6.2.3 Octal System

It was observed that as decimal numbers become larger, the binary number takes up more and more digits, **Octal** System can reduce the digits as it has eight digits 0 to 7. This means that octal has a base of 8 and each digit represents a power of 8.

6.2.4 Hexadecimal System

Hexadecimal system has base of 16 and has 16 digits out of which ten are numbers from 0 to 9; and remaining six are first 6 letters of the alphabet. Hexadecimal codes are used quite often to represent binary numbers. Table 6.3 shows the comparison of binary, decimal and hexadecimal numbers.

Table 6.3 Comparison of binary, decimal and hexadecimal numbers

Binary	Decimal	Hexadecimal
0000	0	0
0001	1	1
0010	2	2
0011	3	3
0100	4	4
0101	5	5
0110	6	6
0111	7	7
1000	8	8
1001	9	9
1010	10	A
1011	11	B
1100	12	C
1101	13	D
1110	14	E
1111	15	F

Groups of four digits can be represented by single digit using hexadecimal base. For example:

$$(1001\ 1100\ 0100)_2 = 9C4_{16}$$
$$(1011\ 0001\ 1010)_2 = B_1A_{16}$$

Decimal equivalent of $9C4_{16}$ is given as:
$$9C4_{16} = (9 \times 16^2) + (C \times 16^1) + (4 \times 16^0)$$
$$= (9 \times 256) + (12 \times 16) + (4 \times 1)$$
$$= (2304 + 192 + 4)_{10}$$
$$= (2500)_{10}.$$

6.3 CONVERSION OF BASES

6.3.1 Decimal to Binary

Successive division of decimal number by 2 is done. The quotient and remainders are noted till the completion of division process. The remainders give the binary number. The first remainder is LSB and the last remainder is MSB. Conversion of decimal number 29 into its binary equivalent is as follows:

2	29
2	14 →1 →LSB
2	7 →0
2	3 →1
2	1 →1 → MSB

Hence, $29_{10} = (11101)_2$

6.3.2 Binary to Decimal

Use weight of a bit = n^{th} bit × 2^{n-1} and add to get decimal number. For example:
$$(11101)^2 = 1 \times 2^4 + 1 \times 2^3 + 1 \times 2^2 + 0 \times 2^1 + 1 \times 2^0$$
$$= 16 + 8 + 4 + 0 + 1 = 29_{10}$$

6.3.3 Fractional Decimal Number to Binary

It is done by successive multiplication of 2 and carry of the number after decimal is recorded. The process is continued until a fractionless decimal number is reached. For example:

Conversion of 0.683 into binary is as follows:

           Carry

    0.683 × 2 = 1.366 1 MSB
    0.366 × 2 = 0.732 0
    0.732 × 2 = 1.464 1
    0.464 × 2 = 0.928 0

$0.928 \times 2 = 1.856 \quad 1$

$0.856 \times 2 = 1.712 \quad 1$

$0.712 \times 2 = 1.424 \quad 1$

$0.424 \times 2 = 0.848 \quad 0$

$0.848 \times 2 = 1.696 \quad 1$

$0.696 \times 2 = 1.392 \quad 1 \quad$ LSB

Hence, $0.683_{10} = (0.1010111011)_2$

6.3.4 Fractional Binary to Decimal

The weightage followed is $2^{-1}, 2^{-2}, 2^{-4}, \ldots 2^{-n}$

For example:

$$(0.11001)_2 = (1 \times 2^{-1} + 1 \times 2^{-2} + 0 \times 2^{-3} + 0 \times 2^{-4} + 1 \times 2^{-5})_{10}$$

$$= \left(\frac{1}{2} + \frac{1}{4} + 0 + 0 + \frac{1}{32}\right)_{10}$$

$$= (0.5 + 0.25 + 0.3125)_{10}$$

or $\quad (0.11001)_2 = (0.0625)_{10}$

6.3.5 Octal to Decimal

Multiply each digit of octal number to the subsequent powers of eight.

For example: $\quad (36)_8 = (3 \times 8^1 + 6 \times 8^0)_{10}$

$\quad\quad\quad\quad\quad\quad = (24 + 6)_{10}$

or $\quad (36)_8 = (30)_{10}$

6.3.6 Decimal to Octal

Continuous division is done by 8. First remainder is LSD and last is MSD.

For example: $(37)_{10}$ Remainder

```
8 | 37
8 |  4    → 5 → LSB
8 |  0    → 4 → MSB
```

∴ $(37)_{10} = (45)_8$

6.3.7 Binary to Octal

The base is $8 = 2^3$, hence, group of 3 bits are formed from right side.

For example: $(101110011)_2 = (101\ 110\ 011)_2$

$\quad\quad\quad\quad\quad\quad\quad\quad = (\ 5 \quad\ 6 \quad\ 3)_8$

6.3.8 Octal to Binary

Each digit of the octal number is converted to its 3-bit binary equivalent.

For example: $(43)_8 = (100\ \ 011)_2$

or $(43)_8 = (100011)_2$

6.3.9 Hexadecimal to Decimal

Multiply the digits to the 16 with its corresponding power.

For example:
$$(8A)_{16} = (8 \times 16^1 + A \times 16^0)_{10}$$
$$= (128 + 10 \times 1)_{10}$$
$$= (138)_{10}$$

6.3.10 Decimal to Hexadecimal

Successive division by 16 is done. First remainder is LSD and last remainder is MSD.

For example: $(73)_{10}$

```
16 | 73
16 |  4    → 9 → LSB
   |  0    → 4 → MSB
```

Hence, $(73)_{10} = (49)_{16}$

6.3.11 Hexadecimal to Binary

Each digit is converted into its equivalent 4 bit binary equivalent.

For example:
$$(5CAB)_{16} = (\underline{0101}\ \ \underline{1100}\ \ \underline{1010}\ \ \underline{1011})_2$$
$$= (0101110010101011)_2$$

6.3.12 Binary to Hexadecimal

Groups of four bits are made from RHS *i.e.*, LSB and then converted into hexadecimal.

For example:
$$(1000111)_2 = (\underline{1000}\ \ \underline{1101})_{16}$$
$$= (8D)_{16}$$

6.3.13 Hexadecimal to Octal

Convert each digit into its binary equivalent, then binary system number is divided into the groups of three bits starting from RHS, *i.e.*, LSB. Convert these groups into octal.

For example:
$$(4AB)_{16} = (0100\ 1010\ 1011)_2$$
$$= (010\ 010\ 101\ 011)_2$$
$$= (\ 2\ \ 2\ \ 5\ \ 3\ )_8$$
$$= (2253)_8$$

6.4 BINARY CODED DECIMAL (BCD) NUMBERS

Each digit of decimal number is converted into four binary bits and group separation is maintained.

For example:
$$(3)_{10} = (0011)_{BCD}$$
$$(12)_{10} = (0001\ 0010)_{BCD}$$

6.5 BINARY ADDITION

Digits are added from LHS, *i.e.*, LSB and carry is taken to RHS for addition.

For example: $(11001)_2 + (10010)_2$

```
              ← carry
    1 1 0 0 1
  + 1 0 0 1 0
  -----------
  1 0 1 0 1 1
```

Hence, $(11001)_2 + (10010)_2 = (101011)_2$

6.6 BINARY SUBTRACTION

Digits are subtracted from LHS *i.e.*, LSB and digits are borrowed from RHS if needed.

For example: $(1101)_2 - (1010)_2$

```
  Borrow→
    1 1 0 1
  - 1 0 1 0
  ---------
    0 0 0 1
```

Hence, $(1101)_2 - (1010)_2 - (0001)_2$

6.7 BOOLEAN ALGEBRA

6.7.1 Basics

De Morgan related **Boolean** with **Algebra**. George Boole constructed an Algebra known as Boolean Algebra. **Boolean Algebra** implements operations of a system of logic required for digital circuits based on ON-OFF/TRUE-FALSE/HIGH-LOW/ 1-0 bistates. Boolean statements may take the form of algebric equations, logic block diagrams, or truth tables as:

Logic variables may have only two values 0 or 1.

Logic Operators of Boolean algebra are:

AND = and (.)

OR = or (+)

NOT = not (−)

XOR = exclusive or $\oplus$

The outputs of any algebraic statements are represented by truth table giving the output values in 0 or 1. Various operator truth tables are as follows:

(i) Truth table for logical operator AND (0)

Inputs		Outputs
A	B	C = A . B
0	0	0
0	1	0
1	0	0
1	1	1

(ii) Truth table for logical operator OR (+)

Inputs		Outputs
A	B	C = A + B
0	0	0
0	1	1
1	0	1
1	1	1

(iii) Truth table for logical operator NOT (−)

Input	Output
A	$\overline{A}$
0	1
0	0

6.8 BOOLEAN ALGEBRA THEOREMS TABLE

S.No.	Name	Theorem
1.	Cummulative Law	$A + B = B + A$
2.	Associative Law	$(A + B) + C = A + (B + C)$ $(A \cdot B) \cdot C = A \cdot (B \cdot C)$
3.	Distributive Law	$A \cdot (B + C) = A \cdot B + A \cdot C$ $A + (B \cdot C) = (A + B) \cdot (A + C)$
4.	Identity Law	$A + A = A$ $A \cdot A = A$
5.	Negation Law	$\overline{\overline{A}} = A$
6.	Redundancy Law	$A + A \cdot B = A$ $A \cdot (A + B) = A$
7.	Boolean Postulates	$0 + A = A$ $1 \cdot A = A$ $0 \cdot A = 0$ $\overline{A} + A = 1$ $\overline{A} \cdot A = 0$ $A + \overline{A} \cdot B = A + B$ $A \cdot (\overline{A} + B) = A \cdot B$
8.	De Morgan's Law	$\overline{A + B} = \overline{A} \cdot \overline{B}$ $\overline{A \cdot B} = \overline{A} + \overline{B}$

6.9 LOGIC GATES AND UNIVERSAL GATES

The elements of digital circuits which implement the switching logic are known as digital **logic gates**, and their symbols are given in the table which follows. Circuit diagrams which include these symbols are known as logic block diagrams.

S.No.	Logic gate	Symbols
1.	AND	Inputs A, B, C → AND gate → $A.B.C$
2.	OR	Inputs A, B, C → OR gate → $A+B+C$
3.	XOR	Inputs A, B → XOR gate → $A.\bar{B}+\bar{A}.B$, $A \oplus B$
4.	NOT	Input A → NOT gate → $\bar{A}$
5.	NAND	Inputs A, B, C → NAND gate → $\overline{A.B.C}$
6.	NOR	Inputs A, B, C → NOR gate → $\overline{A+B+C}$
7.	XNOR	Inputs A, B → XNOR gate → $A.B+\bar{A}.\bar{B}$, $A \odot B$

AND, OR and NOR gates are known as basic gates as any logic expression can be realised using these gates together. The gates which can create any desirable gates are known as **Universal gates** which are NAND and NOR gates.

AND, OR, NOT, *etc.* can be created using only NAND or NOR gates. Thus, any boolean logic statement can be written by using either NAND or NOR gates only.

A general approach in digital circuit design is to write a logic statement or function describing the required process without regard to complexity. Subsequently, logic statement is simplified to allow implementation with a minimum number of logic gates. One of the methods of simplification is by manipulation of the function using theorems until an equivalent is found which requires a minimum combinations of logic gates.

6.10 CANONICAL FORMS

A binary bit has two states—normal state and complementary state. A literal is primed or unprimed variable and each designates an input to a logic gate in a given function. Suppose, there are two variables A and B which are fed to an AND gate. As each variable can have two values, as such possible combinations will be four i.e. $\bar{A}\bar{B}$, $\bar{A}B$, $A\bar{B}$, AB and each of these AND terms is known as minterm or a standard product. Thus, n variables will have 2^n possible minterms.

Similarly, an OR term product of n variables will have maxterm or standard sum. Thus, canonical forms can be of two types, namely Sum of Products (**SOP**) or Product of Sums (**POS**).

The following table show the possible combinations of two variables A and B. It can be noted that each miniterm is the complement of its corresponding maxterm and *vice-versa*. A Boolean expression can be written algebraically from the truth table by expressing each combination of the variables which give a 1 in function by a minterm and then using OR operator for all of the maxterms.

Table for SOP type Canonical Form

Input		Minterms	Output as given
A	B		Y
0	0	$\bar{A}\,\bar{B}$	1
0	1	$\bar{A}\,B$	1
1	0	$A\,\bar{B}$	1
1	1	AB	1

Addition of all the miniterms of high (1) gives the SOP type canonical form:

$$F(A, B) = \bar{A}\bar{B} + \bar{A}B + A\bar{B} + AB$$

Table for POS type Canonical Form

Input		Minterms	Output as given
A	B	Y	
0	0	$A + B$	1
0	1	$A + \bar{B}$	0
1	0	$\bar{A} + B$	0
1	1	$\bar{A} + \bar{B}$	1

Multiplication of all the maxterms of low (0) gives the POS type canonical Form:

$$F(A, B) = (A + \bar{B})(\bar{A} + B)$$

6.11 K-MAP

Karnaugh-map (**K-map**) or witch diagram is a graphical representation of fundamental products in a truth table. This method requires drawing number of squares in a rectangle or squares. Each square represents a minterm. Number of variables decides the number of squares. In case of n variables, the number of squares will be 2^n.

(*i*) K-map for two variables

K-map of two variables will have $2^2 = 4$ squares. There is one square corresponding to each of the possible combination of variables *i.e.*, minterms. K-map in this case is as follows:

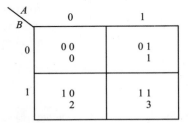
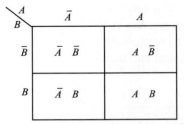

(*ii*) K-map for three variables

K-map of three variables will have $2^3 = 8$ squares. There is one square corresponding to each of the possible combination of variables *i.e.* minterms. K-map in this case is as follows:

AB\C	$\bar{A}\bar{B}$	$\bar{A}B$	AB	$A\bar{B}$
$\bar{C}$	$\bar{A}\bar{B}\bar{C}$	$\bar{A}B\bar{C}$	$AB\bar{C}$	$A\bar{B}\bar{C}$
C	$\bar{A}\bar{B}C$	$\bar{A}BC$	ABC	$A\bar{B}C$

AB\C	00	01	11	10
0	000 0	010 2	110 6	100 4
1	001 1	011 3	111 7	101 5

(*iii*) K-map for four variables

K-map of four variables will have $2^4 = 16$ squares. There is one square corresponding to each of the possible combinations of variables *i.e.*, miniterms. K-map in this case is as follows:

CD\AB	$\bar{A}\bar{B}$	$\bar{A}B$	AB	$A\bar{B}$
$\bar{C}\bar{D}$	$\bar{A}\bar{B}\bar{C}\bar{D}$	$\bar{A}B\bar{C}\bar{D}$	$AB\bar{C}\bar{D}$	$A\bar{B}\bar{C}\bar{D}$
$\bar{C}D$	$\bar{A}\bar{B}\bar{C}D$	$\bar{A}B\bar{C}D$	$AB\bar{C}D$	$A\bar{B}\bar{C}D$
CD	$\bar{A}\bar{B}CD$	$\bar{A}BCD$	$ABCD$	$A\bar{B}CD$
$C\bar{D}$	$\bar{A}\bar{B}C\bar{D}$	$\bar{A}BC\bar{D}$	$ABC\bar{D}$	$A\bar{B}C\bar{D}$

CD\AB	00	01	11	10
00	0000 0	0100 4	1100 12	1000 8
01	0001 1	0101 5	1101 13	1001 9
11	0011 3	0111 7	1111 15	1011 11
10	0010 2	0110 6	1110 14	1010 10

6.12 SIMPLIFICATION OF BOOLEAN EXPRESSION USING K-MAP

Suppose $F = \bar{X}\bar{Y} + X\bar{Y}$ is to be simplified using K-map. K-map is as follows:

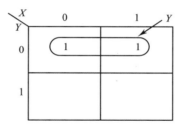

The adjacent two squares showing 1 are grouped together, *i.e.*, encircled. Y is common to both squares. Hence, simplified expression is $F = \bar{Y}$

Now, consider an example with three variables.

Suppose expression $F = \bar{X}\bar{Y}\bar{Z} + \bar{X}Y\bar{Z} + \bar{X}\bar{Y}Z$ is given. K-map is as follows:

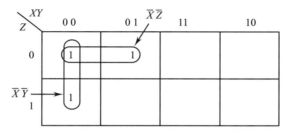

In this case also, two squares, each showing 1 are grouped together *i.e.*, encircled. Hence, simplified expression is $F = \bar{X}\bar{Z} + \bar{X}\bar{Y}$

Further, consider example with four variables

Suppose expression $F = \overline{P}Q\overline{R}S + PQ\overline{R}S + \overline{P}QRS + PQRS + \overline{P}\overline{Q}\overline{R}\overline{S}$

K-map is as follows:

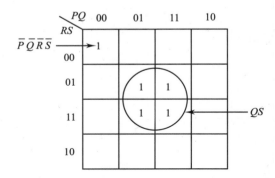

Four squares with 1 are encircled. Column-wise variable Q is common and row-wise S is common. Hence, adjacent four squares give QS after simplification. Moreover, the square representing $\overline{P}\overline{Q}\overline{R}\overline{S}$ is alone, therefore, it is encircled by itself.

Thus, $\qquad F = QS + \overline{P}\overline{Q}\overline{R}\overline{S}$

6.13 SIMPLIFICATION IN SUM OF PRODUCT (SOP) FORM

In this method, groups of o-squares are made and subsequently a sum of products of complementary function is obtained.

Consider expression $\qquad F = (X + Y + \overline{Z})(X + \overline{Y} + \overline{Z})(X + Y + Z)$

or $\qquad F(X, Y, Z) = \pi(0, 1, 3)$

The K-map is as follows:

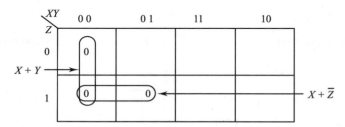

It is found that two vertical o-squares have common variables X and Y with o-logic. There is no common variable from rowside, therefore, the sum is $(X + Y)$. Two horizontal o-squares have Z common with Logic 1, therefore, its complement is taken. From column side, variable X with o-logic is common, therefore, X is taken. The sum is $X + \overline{Z}$. Thus, simplified expression is:

$$F(X, Y, Z) = (X + Y)(X + \overline{Z})$$

SOLVED EXAMPLES

Example 6.1 Convert the following binary numbers into decimal.

(i) 101.01   (ii) 10101.0101

Solution: (i) $(101.01)_2 \to (\ )_{10}$

$$101.01 = 1 \times 2^2 + 0 \times 2^1 + 1 \times 2^0 + 0 \times 2^{-1} + 1 \times 2^{-2}$$

$$= 4 + 0 + 1 + 0 + \frac{1}{4}$$

$$= 5 + \frac{1}{4} = 5.25$$

Therefore, we have

$$(101.01)_2 = (5.25)_{10}$$

(ii) $(10101 \cdot 0101)_2 \to (\ )_{10}$

$$10101.0101 = 1 \times 2^4 + 0 + 2^3 + 1 \times 2^2 + 0 \times 2^1 + 1 \times 2^0$$
$$+ 0 \times 2^{-1} + 1 \times 2^{-2} + 0 \times 2^{-3} + 1 \times 2^{-4}$$

$$= 16 + 0 + 4 + 0 + 1 + 0 + \frac{1}{4} + 0 + \frac{1}{16}$$

$$= 21.3125$$

Therefore, we have

$$(10101 \cdot 0101)_2 = (21.3125)_{10}.$$

Example 6.2 Convert $(4320)_{10}$ into binary number system.

Solution: $(4320)_{10} \to (\ )_2$

2	4320	
2	2160	→0
2	1080	→0
2	540	→0
2	270	→0
2	135	→0
2	67	→1
2	33	→1
2	16	→1
2	8	→0
2	4	→0
2	2	→0
2	1	→1
	0	

(Reading from bottom to the top)

Therefore, we obtain $(4320)_{10}$
$= (100011100000)_2$

Example 6.3 Convert $(3451)_{10}$ into binary number system.

Solution:

$$\begin{array}{r|l}
2 & 3451 \\
\hline
2 & 1725 \to 1 \\
\hline
2 & 862 \to 1 \\
\hline
2 & 431 \to 0 \\
\hline
2 & 215 \to 1 \\
\hline
2 & 107 \to 1 \\
\hline
2 & 53 \to 1 \\
\hline
2 & 26 \to 1 \\
\hline
2 & 13 \to 0 \\
\hline
2 & 6 \to 1 \\
\hline
2 & 3 \to 0 \\
\hline
2 & 1 \to 1 \\
\hline
& 0 \to 1
\end{array}$$

Reading from bottom to the top.

Therefore, we obtain

$$(3451)_{10} = (110101111011)_2.$$

Example 6.4 Convert the decimal number $(250.5)_{10}$ to base 3, base 4, base 7 and base 16.

Solution: Given Number = $(250.5)_{10}$

For Base 3

$$\begin{array}{r|l}
3 & 250 \\
\hline
3 & 83 \to 1 \\
\hline
3 & 27 \to 2 \\
\hline
3 & 9 \to 0 \\
\hline
3 & 3 \to 0 \\
\hline
3 & 1 \to 0 \\
\hline
& 0 \to 1
\end{array}$$

$$\begin{array}{l}
0.5 \\
\times 3 \\
\hline
1.5 \to 0.5 \\
\downarrow \quad \times 3 \\
1 \quad\quad 1.5
\end{array}$$

or $(250.5)_{10} = (100021.11)_3$

For Base 4

$$\begin{array}{r|l}
4 & 250 \\
\hline
4 & 62 \to 2 \\
\hline
4 & 15 \to 2 \\
\hline
4 & 3 \to 3 \\
\hline
& 0 \to 3
\end{array}$$

$$\begin{array}{l}
0.5 \\
\times 4 \\
\hline
2.0 \\
\downarrow \\
2
\end{array}$$

or $(250.5)_{10} = (3322.2)_4$

For Base 7

```
7 | 250
7 |  35 → 5
7 |   5 → 0
  |   0 → 5
```

```
  0 . 5
  × 7
  3. 5  → 0.5
  ↓       × 7
  3       3. 5
          ↓
          3
```

or $(250.5)_{10} = (505.33)_7$

For Base 16

```
16 | 250
16 |  15 → A
   |   0   F  ↑
```

```
  0 . 5
  × 16
  8. 0
  ↓
  8
```

or $(250.5)_{10} = (FA.8)_{16}$.

Example 6.5 Convert $(1201102)_3 = (\ )_{10}$

Solution: $(1201102)_3 = 1 \times 3^6 + 2 \times 3^5 + 0 \times 3^4 + 1 \times 3^3 + 1 \times 3^2 + 0 \times 3^1 + 2 \times 3^0$

$= 729 + 486 + 0 + 27 + 9 + 0 + 2$

$(1201102)_3 = (1253)_{10}$.

Example 6.6 Add the following binary numbers.

11010101 and 1101101

Solution: We have

```
  1 1 1 1 1  1    ← carry
    1 1 0 1 0 1 0 1
      1 1 0 1 1 0 1
    1 0 1 0 0 0 0 1 0
```

Example 6.7 Add $(10111010)_2$ and $(101001)_2$.

Solution: We have

```
      1 1 1         ← carry
    1 0 1 1 1 0 1 0
        1 0 1 0 0 1
    1 1 1 0 0 0 1 1
```

Example 6.8 Add the following without changing the base.

(i) $(734)_8 + (444)_8$

(ii) $(432)_5 - (124)_5$

(iii) $(A1D)_{16} + (99F)_{16}$

Solution: (i) $(734)_8 + (444)_8$

Here, we have

$$\begin{array}{r} 1\,1 \quad \leftarrow \text{carry} \\ (7\ 3\ 4)_8 \\ +\ (4\ 4\ 4)_8 \\ \hline (1\ 4\ 0\ 0)_8 \end{array}$$

Therefore, we write

$$(734)_8 + (444)_8 = (1400)_8$$

(ii) $(432)_5 - (124)_5$

Thus,

$$\begin{array}{r} (432)_5 \\ -\ (124)_5 \\ \hline (303)_5 \end{array}$$

(iii) $(A1D)_{16} + (99F)_{16}$

$$\begin{array}{r} 1 \\ (A1D)_{16} \\ +\ (99F)_{16} \\ \hline (13\,B\,C)_{16} \end{array}$$

Example 6.9 Write the sum of $(23.53)_{10}$ and $(23.53)_8$ in decimal.

Solution: Firstly, let us convert $(23.53)_8$ in decimal and then add it with $(23.53)_{10}$.

Thus, we have,

$$(23.53)_8 = 2 \times 8^1 + 3 \times 8^0 + 5 \times 8^{-1} + 3 \times 8^{-2}$$

$$= 16 + 3 + \frac{5}{8} + \frac{3}{64}$$

$$= 16 + 3 + 0.625 + 0.046875$$

$$(23.53)_8 = (19.671875)_{10}$$

Therefore,

$$\begin{array}{r} (23.53)_{10} \\ (19.671875)_{10} \\ \hline (43.201875)_{10} \end{array}$$

Example 6.10 Convert the following hexadecimals into decimals.

(i) A 13 B (ii) 7C A 3 (iii) 7F D6

Solution: (i) $(A\ 13\ B)_{16} \rightarrow (\ )_{10}$

We have

$$\begin{array}{cccc} A & 1 & 3 & B \\ \downarrow & \downarrow & \downarrow & \downarrow \\ 10 & 1 & 3 & 11 \end{array}$$

$$= 10 \times 16^3 + 1 \times 16^2 + 3 \times 16^1 + 11 \times 16^0$$
$$= 40960 + 256 + 48 + 11$$
$$= (41275)_{10}$$

Therefore, we obtain

$$(A\ 13\ B)_{16} = (41275)_{10}$$

(ii) $(7\ C\ A\ 3)_{16} \rightarrow (\ )_{10}$

We have

$$\begin{array}{cccc} 7 & C & A & 3 \\ \downarrow & \downarrow & \downarrow & \downarrow \\ 7 & 12 & 10 & 3 \end{array}$$

$$= 7 \times 16^3 + 12 \times 16^2 + 10 + 16^1 + 3 \times 16^0$$
$$= 28672 + 3072 + 160 + 3$$
$$(7CA3)_{16} = (31907)_{10}$$

(iii) $(7FD6)_{16} \rightarrow (\ )_{10}$

We have

$$\begin{array}{cccc} 7 & F & D & 6 \\ \downarrow & \downarrow & \downarrow & \downarrow \\ 7 & 15 & 13 & 6 \end{array}$$

$$= 7 \times 16^3 + 15 \times 16^2 + 13 \times 16^1 + 6 \times 16^0$$
$$= 28672 + 3840 + 208 + 6$$
$$= 32726$$

Therefore, we have $(7FD6)_{16} = (32726)_{10}$.

Example 6.11 Obtain the following conversion.

(i) $(23 \cdot AB)_{16} \rightarrow (\ )_2$

Solution: We have

$$\begin{array}{cccc} (23 \cdot AB)_{16} = & 2 & 3. & A & B \\ & \downarrow & \downarrow & \downarrow & \downarrow \\ & 0010 & 0011 & 1010 & 1011 \end{array}$$

Therefore, we obtain

$$(23 \cdot AB)_{16} = (00100011.10101011)_2.$$

Example 6.12 Convert the hexadecimal number $(1CD \cdot 2A)_{16}$ to binary.

Solution: $(1CD \cdot 2A)_{16} \rightarrow (\quad)_2$

We have

1	$C_{(12)}$	$D_{(13)}$	.2	$A_{(10)}$
↓	↓	↓	↓	↓
0001	1100	1101	0010	1010

Therefore, we obtain
$$(1CD \cdot 2A)_{16} = (000111001101 \cdot 00101010)_2.$$

Example 6.13 Add the following Hexadecimal Number.

(i) 93 + DE

(ii) ABCD + EF 12

Solution: (i) 93 + DE

We have

```
           11  ← carry
        ( 9   3)₁₆
     +  ( D   E)₁₆
        ─────────
         (17  1)₁₆
```

(ii) $(ABCD)_{16} + (EF12)_{16}$

```
         11               ← carry
         A    B    C    D
         E    F    1    2
        ────────────────────
        (19   A    D    F)₁₆
```

Example 6.14 Convert the octal number $(1745.246)_8$ into hexadecimal number.

Solution: $(1745.246) = (\quad)_{16}$

We have

1	7	4	5	.	2	4	6
↓	↓	↓	↓	↓	↓	↓	↓
001	111	100	101	.	010	100	110

So, $(1745.346)8 = (001111\ 100101 \cdot 010100110)_2$

$$= \underbrace{0011}_{3}\underbrace{1110}_{E}\underbrace{0101}_{5} \cdot \underbrace{0101}_{5}\underbrace{0011}_{3}\underbrace{0000}_{0}$$

$$= (3E5.530)_{16}.$$

Example 6.15 Simplify

$$F(a, b, c) = \overline{a}bc + b\overline{c} + ab\overline{c} + a\overline{b}c \text{ using k-map.}$$

Solution: Here we have

$$F(a, b, c) = \overline{a}bc + b\overline{c} + ab\overline{c} + a\overline{b}c$$

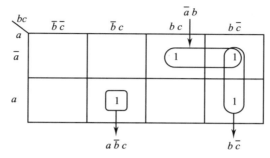

So, $\overline{a}bc + b\overline{c} + ab\overline{c} + a\overline{b}c = \overline{a}b + a\overline{b}c + b\overline{c}$
Thus, we have

$$F(a, b, c) = a\overline{b}c + \overline{a}b + b\overline{c}.$$

Example 6.16 Obtain 1's and 2's complement of 1010101, 0111000

Solution: (*i*)   Binary No. = 1010101
               1's complement = 0101010
               2's complement = 1's complement + 1
                               = 0101010 + 1
                                 = 0101011.

(*ii*)   Binary No. = 0111000
               1's complement = 1000111
               2's complement = 1's complement + 1
                               = 1000111 + 1
                                 = 1001000.

Example 6.17 Convert 2 AC5 · D to Octal

Solution: 2AC5 · D to Octal

	2	A	C	5	·	D (Hex)
	↓	↓	↓	↓		↓
	0010	1010	1100	0101	·	1101(Binary)
	010	101	011	000 101	·	110100
	2	5	3	05	·	64 (Octal)

or   **(2 AC5 · D)16 ≡ (25305 · 64)$_8$.**

Example 6.18 Using 10's complement, subtract 72532 – 3250. Also perform the operation using 9's complement.

Solution: Let $\quad$ M = 72532

$$\& \text{ N} = 3250 = 03250$$

$$\text{M} = 72532$$

10's complement of $\quad$ N = 96750

$$169282$$

neglecting end carry

Sum → 69282

Using by 9's complement

9's complement of N = 96749

$$\text{M} = 72532$$
$$\text{N} = 96749 \qquad\qquad 9's$$
$$\overline{169281}$$
$$1$$
$$\overline{69282}$$

Example 6.19 Convert Decimal Number 225.225 to Octal and Hexadecimal.

Solution: Decimal Number = 225.225

(*i*) Decimal to Octal

Decimal	Integer Part	Remainder
8	225	–
	28	1
	3	4
	0	3

Octal equivalent $(225)_{10} = (341)_8$

Fractional Part

$$0.225 \times 8 = 0.800 \text{ with a carry of 1}$$
$$0.800 \times 8 = 0.400 \text{ with a carry of 6}$$
$$0.400 \times 8 = 0.200 \text{ with a carry of 3}$$
$$0.200 \times 8 = 0.600 \text{ with a carry of 1}$$
$$0.600 \times 8 = 0.800 \text{ with a carry of 4}$$

∵ $\quad (0.225)_{10} \equiv (0.16314)_8$

Hence, $\quad (225.225)_{10} \equiv (341.16314)_8$

(*ii*) Decimal to Hexadecimal

Decimal	Integer Part	Remainder
16	225	–
	14	1
	0	14 = E

Hex equivalent ∵ $(225)_{10} = (E1)_{16}$

Fractional Part $0.225 \times 16 = 0.600$ with a carry of 3

$0.600 \times 16 = 0.600$ with a carry of 9

∴ $(0.225)_{10} \equiv (.39)_{16}$

Hence, $(225.225)_{10} \equiv (E\ 1.39)_{16}$.

Example 6.20 Write the dual of the following theorem:

$$A + (B \cdot C) = (A + B) \cdot (A + C)$$

Solution: Given theorem is

$$A + (B \cdot C) = (A + B) \cdot (A + C)$$

To obtain dual of above theorem, replacing '+' by '·', '·' by '+' and complementing 0's and 1's.

Dual $A \cdot (B + C) = (A \cdot B) + A \cdot C$

$A \cdot (B + C) = A \cdot B + A \cdot C.$

Example 6.21 Obtain 9's and 10's compliment of 13579, 09900.

Solution: (*i*)   Decimal Number = 13579

9's complement = 99999

                   13579
                   ―――――
                   86420

10's complement = 9's complement + 1

= 86420 + 1

= 86421

(*ii*)   Decimal No. = 09900

9's Complement = 99999

                   09900
                   ―――――
                   90099

10's complement = 90099 + 1

= 90100

Example 6.22 Simplify the following Boolean equations.

(i) $F = B \cdot (A + B)$

(ii) $F = \overline{A} + \overline{B} + A \cdot B \cdot \overline{C}$

(iii) $F = \overline{A}\,\overline{B}\,\overline{C}\,\overline{D} + \overline{A}\,\overline{B}\,\overline{C} \cdot D$

(iv) $F = \overline{A}C + \overline{A}B + A \cdot \overline{B}C + BC$

Solution:

(i)
$$F = B \cdot (A + B) = B \cdot A + B \cdot B$$
$$F = B \cdot A + B \quad [\because\ x \cdot x = x]$$
$$F = B(A + 1)$$
$$= B \cdot 1 \quad [X + 1 = 1]$$
$$= B$$

(ii)
$$F = \overline{A} + \overline{B} + A \cdot B \cdot \overline{C}$$
$$= \overline{A} \cdot 1 + \overline{B} + A \cdot B \cdot \overline{C}$$
$$= \overline{A}(1 + B\overline{C}) + \overline{B} + A \cdot B\overline{C} \quad [\because\ 1 + x = 1]$$
$$= \overline{A} + \overline{A}B\overline{C} + \overline{B} + AB\overline{C}$$
$$= B\overline{C}(A + \overline{A}) + \overline{A} + \overline{B}$$
$$= B\overline{C} \cdot 1 + \overline{A} + \overline{B} \quad [\because\ x + \overline{x} = 1]$$
$$= B\overline{C} + \overline{A} + \overline{B}$$

(iii)
$$F = \overline{A}\,\overline{B}\,\overline{C}\,\overline{D} + \overline{A}\,\overline{B}\,\overline{C} \cdot D$$
$$= \overline{A}\,\overline{B}\,\overline{C}(D + \overline{D})$$
$$= \overline{A}\,\overline{B}\,\overline{C} \cdot 1$$
$$F = \overline{A}\,\overline{B}\,\overline{C}$$

(iv)
$$F = \overline{A}C + \overline{A}B + A \cdot \overline{B}C + BC$$
$$F = \overline{A}B + C(\overline{A} + A\overline{B} + B)$$
$$= \overline{A}B + C[\overline{A} + A\overline{B} + B(A + \overline{A})]$$
$$= \overline{A}B + C[\overline{A} + A(B + \overline{B}) + \overline{A} \cdot B]$$
$$= \overline{A}B + C[\overline{A} + A \cdot 1 + \overline{A}B]$$
$$= \overline{A}B + C[1 + \overline{A}B] \quad [\because\ A + \overline{A} = 1]$$
$$= \overline{A}B + C \cdot 1 \quad [1 + \overline{A}B = 1]$$
$$F = \overline{A}B + C.$$

Example 6.23 Express the following functions in a sum of minterms and a product of maxterms.

(i) $F(x, y, z) = 1$

(ii) $F(A, B, C, D) = D \cdot (\overline{A} + B) + \overline{B} \cdot D$

Solution: (*i*) $F(x, y, z) = 1$

Since $F(x, y, z)$ equals 1, the minterms are:
$$F(x, y, z) = \bar{x}\bar{y}\bar{z} + \bar{x}\bar{y}z + x\bar{y}\bar{z} + \bar{x}y\bar{z} + \bar{x}yz + x\bar{y}z + xy\bar{z} + xyz$$
$$F(x, y, z) = \Sigma\, m\,(0, 1, 2, 3, 4, 5, 6, 7)$$

From above it is clear that there is no maxiterm in the given expression.

(*ii*) $F(A, B, C, D) = D(\bar{A} + B) + \bar{B}D$

$$F(A, B, C, D) = D \cdot \bar{A} + D \cdot B + \bar{B} \cdot D$$

Now, we will simplify term by term
$$D\bar{A} = D\bar{A}(B + \bar{B})$$
$$= D\bar{A}B + D\bar{A}\bar{B}$$
$$= D\bar{A}B[C + \bar{C}] + D\bar{A}\bar{B}[C + \bar{C}] \qquad [\because\ X + \bar{X} = 1]$$
$$D\bar{A} = D\bar{A}BC + D\bar{A}B\bar{C} + D\bar{A}\bar{B}C + D\bar{A}\bar{B}\bar{C}$$
$$D\bar{A} = \bar{A}BCD + \bar{A}B\bar{C}D + \bar{A}\bar{B}CD + \bar{A}\bar{B}\bar{C}D \qquad \ldots(i)$$

$$DB = D \cdot B \cdot 1 = D \cdot B(A + \bar{A}) = D \cdot BA + D \cdot B \cdot \bar{A}$$
$$= D \cdot B \cdot A[C + \bar{C}] + D \cdot B \cdot \bar{A}$$
$$[C + \bar{C}] = D \times B \times A \times C + D \times B \times A \times \bar{C} + D \times B \times \bar{A} \times C$$
$$+ D \times B \times A \times C \qquad \ldots(ii)$$

$$\bar{B}D = \bar{B}D(A + \bar{A}) = \bar{B}DA + \bar{B}D\bar{A}$$
$$= \bar{B}DA(C + \bar{C}) + \bar{B}D \cdot \bar{A}(C + \bar{C})$$
$$= \bar{B}DAC + \bar{B}DA\bar{C} + \bar{B}D\bar{A}C + \bar{B}D\bar{A}\bar{C}$$
$$= A\bar{B}CD + A\bar{B}\bar{C}D + \bar{A}\bar{B}CD + \bar{A}\bar{B}\bar{C}D \qquad \ldots(iii)$$

Therefore = $D\bar{A} + D \cdot B + \bar{B}D$,

Putting the values from (*i*), (*ii*) & (*iii*)

$$F(A, B, C, D) = \bar{A}BCD + \bar{A}B\bar{C}D + \bar{A}\bar{B}CD + \bar{A}\bar{B}\bar{C}D + ABCD + ABCD$$
$$+ \bar{A}BCD + \bar{A}B\bar{C}D + A\bar{B}CD + A\bar{B}\bar{C}D + \bar{A}\bar{B}CD + \bar{A}\bar{B}\bar{C}D$$

$$F(A, B, C, D) = \bar{A}BCD + \bar{A}B\bar{C}D + \bar{A}\bar{B}CD + \bar{A}\bar{B}\bar{C}D$$
$$+ ABCD + AB\bar{C}D + A\bar{B}CD + A\bar{B}\bar{C}D \qquad [\because\ X + \bar{X} = 1]$$

$$F(A, B, C, D) = m_7 + m_5 + m_3 + m_1 + m_{15} + m_{13} + m_{11} + m_9$$
$$F(A, B, C, D) = \Sigma\, m\,(1, 3, 5, 7, 9, 11, 13, 15)$$

∴ Maxterms are = $\pi\, M\,(0, 2, 4, 6, 8, 10, 12, 14)$.

Example 6.24 Convert the following into other canonical forms:

(a) $f(x, y, z) = \Sigma m (1, 3, 7)$

(b) $f(x, y, z) = \pi m (0, 3, 6, 7)$

Solution: (a) $f(x, y, z) = S m (1, 3, 7)$

∵ Compliment $f'(x, y, z) = \Sigma m (0, 2, 4, 5, 6)$

$$f'(x, y, z) = m_0 + m_2 + m_4 + m_5 + m_6$$

Using De Morgan's Theorem

$$f(x, y, z) = \overline{m_0 + m_2 + m_4 + m_5 + m_6} = \overline{m_0} \cdot \overline{m_2} \cdot \overline{m_4} \cdot \overline{m_5} \cdot \overline{m_6}$$
$$f(x, y, z) = M_0 \cdot M_2 \cdot M_4 \cdot M_5 \cdot M_6$$
$$f(x, y, z) = \pi M (0, 2, 4, 5, 6)$$

(b) $\quad f(x, y, z) = \pi m (0, 3, 6, 7)$

Complement $\quad f'(x, y, z) = \pi m (1, 2, 4, 5)$

$$f'(x, y, z) = m_1 \cdot m_2 \cdot m_4 \cdot m_5$$

Using De Morgan's Theorem

$$f(x, y, z) = \overline{M_1 \cdot M_2 \cdot M_4 \cdot M_5} = \overline{M_1} + \overline{M_2} + \overline{M_4} + \overline{M_5}$$
$$f(x, y, z) = m_1 + m_2 + m_4 + m_5$$
$$f(x, y, z) = \Sigma m (1, 2, 4, 5).$$

Example 6.25 Obtain the simplified expression in sum of product for the following Boolean functions using K-map.

(i) $f = \bar{x} y z + x \bar{y} \bar{z} + x y z + x y \bar{z}$

(ii) $f(x, y, z) = \Sigma m (0, 2, 4, 5, 6)$

Solution:

(i) $f = \bar{x} y z + x \bar{y} \bar{z} + x y z + x y \bar{z}$

K-map representation for this expression is

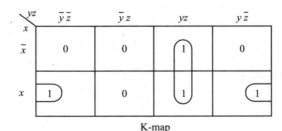

K-map

In the K-map, the right corner and left corner make a square and the common term is $x\bar{z}$. In the middle square, the common term is yz. Hence, combining the above two, the simplified expression is

$$f = x\bar{z} + yz$$

(ii) $\quad f(x, y, z) = \Sigma m (0, 2, 4, 5, 6) = m_0 + m_2 + m_4 + m_5 + m_6$

$\quad f(x, y, z) = \bar{x}\bar{y}\bar{z} + \bar{x} y\bar{z} + x\bar{y}\bar{z} + x\bar{y} z + x y \bar{z}$

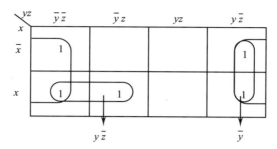

K-map representation for this expression is

In the K-map, combining the four adjacent square in the first and last columns, the common term is $\bar{z}$. The two squares give the common term $x\bar{y}$

Hence, the simplified expression is

$$f(x, y, z) = \bar{z} + x\bar{y}$$

$$f(x, y, z) = \bar{z} + x\bar{y}$$

Example 6.26 Given the following truth table:

x	y	z	f_1	f_2
0	0	0	0	0
0	0	1	1	0
0	1	0	1	0
0	1	1	0	1
1	0	0	1	0
1	0	1	0	1
1	1	0	0	1
1	1	1	1	1

(a) Express f_1 and f_2 in product of maxiterms.
(b) Obtain the simplified function in SOP form using Cap K-map.

Solution: (a) From truth table, we have

$$f_1 = \bar{x}\bar{y}z + \bar{x}y\bar{z} + x\bar{y}\bar{z} + xyz = m_1 + m_2 + m_4 + m_7$$

∵

$$\bar{f_1} = m_0 + m_3 + m_5 + m_6$$

Using De Morgan's theorem

$$f_1 = \overline{m_0 + m_3 + m_5 + m_6}$$

$$f_1 = \bar{m_0} \cdot \bar{m_3} \cdot \bar{m_5} \cdot \bar{m_6}$$

$$f_1 = M_0 \cdot M_3 \cdot M_5 \cdot M_6$$

or

$$f_1 = (x+y+z)\cdot(x+\bar{y}+\bar{z})\cdot(\bar{x}+y+\bar{z})\cdot(\bar{x}+\bar{y}+z)$$

Similarly, form truth table, we have

$$f_2 = \bar{x}yz + x\bar{y}z + xy\bar{z} + xyz$$
$$f_2 = m_3 + m_5 + m_6 + m_7$$
$$\overline{f_2} = m_0 + m_1 + m_2 + m_4$$

Using De Morgan's theorem

$$f_{2'} = \overline{m_0 + m_1 + m_2 + m_4}$$
$$f_2 = \overline{m_0} \cdot \overline{m_1} \cdot \overline{m_2} \cdot \overline{m_4}$$
$$f_2 = M_0 \cdot M_1 \cdot M_2 \cdot M_4$$

From truth table $\quad f_2 = (x+y+z) + (x+y+\bar{z}) \cdot (x+\bar{y}+z) \cdot (\bar{x}+y+z)$

(b) $\quad f_1(x, y, z) = m_1 + m_2 + m_4 + m_7 = \Sigma m (1, 2, 4, 7)$

$$f_1(x, y, z) = \bar{x}\bar{y}z + \bar{x}y\bar{z} + x\bar{y}\bar{z} + xyz$$

K-map representation

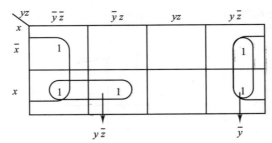

The above K-map cannot be minimised since no pair is possible

∴ $\quad f_1(x, y, z) = \bar{x}\bar{y}z + \bar{x}y\bar{z} + x\bar{y}\bar{z} + xyz$

Similarly $\quad f_2(x, y, z) = m^3 + m^5 + m^6 + m^7$

$$f_2(x, y, z) = \Sigma(3, 5, 6, 7)$$
$$f_2(x, y, z) = \bar{x}yz + x\bar{y}z + xy\bar{z} + xyz$$

K-map representation

	$\bar{y}\bar{z}$	$\bar{y}z$	yz	$y\bar{z}$
$\bar{x}$			1	
x		1	1	1

Two vertical squares reduce to yz

Two horizontal squares reduce to xy

The one separate square is $x\bar{y}z$

Therefore, the simplified expression is

$$f_2(x, y, z) = yz + xy + x\bar{y}z.$$

Example 6.27 Convert the Hex no. (1 CD · 2A) to binary and decimal numbers.

Solution: (*i*) Hex to Binary

Hex No. = 1 CD · 2A

1	C	D	·	2	A	(Hex)
↓	↓	↓		↓	↓	
0001	1100	1101	·	0010	1010	(Binary)

Hence, $(1CD \cdot 2A)_{16} = (0001\ 1100\ 1101 \cdot 00101010)_2$

(*ii*) Hex to Decimal

Hex No. = 1 CD · 2A

Ist we note that C stands for 12 in decimal. D stands for 13 in decimal and A stands for 10 in decimal.

∵ $1CD \cdot 2A = 1 \times 16^2 + C \times 16^1 + D \times 16^0 + 2 \times 16^{-1} + A \times 16^{-2}$

$1CD \cdot 2A = 256 + 12 \times 16 + 13 \times 1 + \dfrac{2}{16} + 10 \times \dfrac{1}{256}$

$= 256 + 192 + 13 + 0.125 + 0.039$

$(1CD \cdot 2A)_{16} \equiv (461 \cdot 164)_{10}$.

Example 6.28 Convert the Binary Number (1011.011) into Octal and Hexadecimal numbers.

Solution: (i) Binary to Octal:

Given Binary number = 1011 · 011

10	11	·	011
001	011	·	011
1	3	·	3

∵ $(1011.011)_2 = (13.3)_8$

(*ii*) Binary to Hex

Given Binary number = 1011 · 011

10 11	·	01 10
B		6

$(1011.011)_2 = (B.6)_{16}$.

Example 6.29. Add and subtract without converting the following two numbers.

$(7571)_8$ and $(4176)_8$

Solution: The given Nas are $(7571)_8$ and $(4176)_8$

Addition

           1 1 ← carry
         $(7\ 5\ 7\ 1)_8$
         $(4\ 1\ 7\ 6)_8$
         $(1\ 3\ 7\ 6\ 7)_8$

Therefore, $(7571)_8 + (4176)_8 = (13767)_8$
Subtraction: $\quad (7\ 5\ 7\ 1)_8$
$\quad\quad\quad\quad\quad\quad\quad (4\ 1\ 7\ 6)_8$
$\quad\quad\quad\quad\quad\quad\quad \overline{(3\ 3\ 7\ 3)_8}$

Therefore, $(7571)_8 - (4176)_8 = (3373)_8$.

Example 6.30 Add and subtract the following 2 numbers without converting into decimal number.

$$(432)_5 \quad \text{and} \quad (013)_5$$

Solution: Since base is 5, therefore on simple adding if no. exceeds 4, then, it will get converted into base 5.

Addition $\quad\quad\quad\quad\quad\quad 4\ 3\ 2$
$\quad\quad\quad\quad\quad\quad\quad\quad\quad 0\ 1\ 3$
$\quad\quad\quad\quad\quad\quad\quad\quad \overline{1\ 0\ 0\ 0}$

Therefore, $(432)_5 + (013)_5 = (1000)_5$

Subtraction $\quad\quad\quad\quad\quad 4\ 3\ 2$
$\quad\quad\quad\quad\quad\quad\quad\quad\quad - 0\ 1\ 3$
$\quad\quad\quad\quad\quad\quad\quad\quad \overline{\quad 4\ 1\ 4}$

Therefore, $(432)_5 - (013)_5 = (414)_5$.

Example 6.31 Convert the following numbers as indicated.

(i) $(IBE)_{16} = (\quad)_8$
(ii) $(676)_8 = (\quad)_2$
(iii) $(321)_4 = (\quad)_{10}$

Solution: (i) $(1BE)_{16}$

$\quad\quad 1 \quad\quad\quad B \quad\quad\quad E \quad\quad\quad$ Hexadecimal
$\quad\quad \underbrace{0001} \quad \underbrace{1011} \quad \underbrace{1110}$
$\quad\quad\quad\quad\quad\quad\quad\quad\quad\quad\quad\quad\quad\quad\quad$ Binary

$\quad\quad \underbrace{000}_{0} \quad \underbrace{110}_{6} \quad \underbrace{111}_{7} \quad \underbrace{110}_{6}$
$\quad\quad\quad\quad\quad\quad\quad\quad\quad\quad\quad\quad\quad\quad\quad$ Octal

So, $\quad\quad\quad (IBE)_{16} = (676)_8$
Hence $\quad\quad (1BE)_{16} \equiv (0676)_8$

(ii) $\quad\quad\quad (676)8 \equiv \quad 6 \quad\quad 7 \quad\quad 6$
$\quad\quad\quad\quad\quad\quad\quad\quad\quad 110 \quad 111 \quad 110$
$\quad\quad\quad\quad\quad (676)_8 \equiv (110111110)_2$

(iii) $\quad\quad\quad (321)_4 = 3 \times 4^2 + 2 \times 4^1 + 1 \times 4^0$
$\quad\quad\quad\quad\quad\quad\quad = 48 + 8 + 1$

Thus, $\quad\quad\quad (321)_4 = (57)_{10}$.

Example 6.32 How an exclusive NOR gate can be obtained using NAND gate only? Sketch the diag.

Solution: An exclusive NOR gate is represented as

Here
$$Y = A \odot B$$
$$Y = A \cdot B + \overline{A} \cdot \overline{B}$$

For EX-NOR gate, the truth table representation

A	B	$Y = A \cdot B + \overline{A} \cdot \overline{B}$
0	0	1
0	1	0
1	0	0
1	1	1

Representation of Ex-NOR using NAND gate only

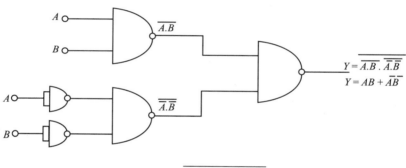

Now
$$Y = \overline{(A \cdot B) \cdot (\overline{A} \cdot \overline{B})}$$
$$Y = \overline{\overline{A \cdot B}} + \overline{(\overline{A} \cdot \overline{B})}$$

Using De Morgan's theorem
$$Y = A \cdot B + \overline{A} \cdot \overline{B}$$
$$Y = A \odot B$$
$$Y = \Sigma\text{-NOR gate.}$$

Example 6.33 Simplify $F(ABCD) = ABC + BCD + A\overline{C}D + A\overline{B} + A$ by using K-map.

Solution: The given expression is $F(ABCD) = ABC + BCD + A\overline{C}D + A\overline{B} + A$. The K-map of the given expression is ahead:

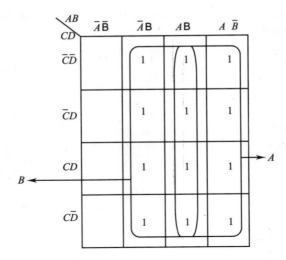

Hence, the simplified function is given as $F(ABCD) = A + B$.

Example 6.34 Simplify $F(a, b, c) = a\bar{b} + \bar{b}c + \bar{c}a$ using K-map.

Solution: The given function is $F = a\bar{b} + \bar{b}c + \bar{c}a$

Since this is a SOP form consisting of a 3 literals (a, b, c) therefore, it can be converted into standard SOP form as under:

$$F = a\bar{b} + \bar{b}c + \bar{c}a$$

$$F = (c+\bar{c})\bar{a}b + \bar{b}c(a+\bar{a}) + \bar{c}a(b+\bar{b}) \qquad [x + \bar{x} = 1]$$

$$F = \bar{a}bc + \bar{a}b\bar{c} + a\bar{b}c + \bar{a}\bar{b}c + ab\bar{c} + a\bar{b}\bar{c}$$

$$F = 011 + 010 + 101 + 001 + 110 + 100$$

$$F = 3 \quad\quad 2 \quad\quad 5 \quad\quad 1 \quad\quad 6 \quad\quad 4$$

$$F = m_3 + m_2 + m_5 + m_1 + m_6 + m_4$$

$$F = \Sigma\, m(3, 2, 5, 1, 6, 4)$$

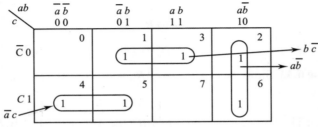

Therefore, simplified form is

$$F(a, b, c) = \bar{a}c + b\bar{c} + a\bar{b}.$$

Example 6.35 Convert 24_{10} to binary.

Solution: The highest power of 2 not greater than 24 is $2^4 = 16$.

Take 16 from 24 to leave 8.

The highest power of 2 not greater than 8 is $2^3 = 8$.

Take 8 from 8 to leave 0.

The binary for 24_{10} is

$2^4 + 2^3$ equivalent to 11000

Thus, $24_{10} = 11000_2$

This approach can easily lead to mistake, principally because we can overlook the powers which have zero digits. An alternative which is more reliable is repeatedly to divide the decimal number by 2, the remainder indicating the appropriate binary digit. This is known as the repeated division by 2 method.

Example 6.36 Convert 24_{10} to binary.

Solution:

2	24	Remainder 0	0	Least significant digit (LSB)
2	12	Remainder 0	0	
2	6	Remainder 0	0	
2	3	Remainder 1	1	
2	1	Remainder 1	1	Most significant digit (MSB)
	0			

Thus, $24_{10} = 11000_2$.

Example 6.37 Convert 33_{10} to binary.

Solution:

2	33	Remainder 1	1	(LSB)
2	16	Remainder 0	0	
2	8	Remainder 0	0	
2	4	Remainder 0	0	
2	2	Remainder 0	0	(MSB)
2	1	Remainder 0	1	
	0			

Thus, $33_{10} = 100001_2$.

Example 6.38 Add the binary number 1010 and 80110
Solution:

$$\begin{array}{r} 1\ 1 \quad \leftarrow \text{Carry} \\ 1\ 0\ 1\ 0 \\ 0\ 1\ 1\ 0 \\ \hline 1\ 0\ 0\ 0\ 0_2 \end{array}$$

Example 6.39 Add the decimal number 19 and 9 by binary means.
Solution:

$$\begin{array}{rl} 1\ 1 & \leftarrow \text{Carry} \\ 1\ 0\ 0\ 1\ 1 & 19 \\ 0\ 1\ 0\ 0\ 1 & 9 \\ \hline 1\ 1\ 1\ 0\ 0 & \equiv 28_{10} \end{array}$$

Example 6.40 Add the decimal number 79 and 31 by binary means.
Solution:

$$\begin{array}{rl} 1\ 1\ 1\ 1\ 1 & \leftarrow \text{Carry} \\ 1\ 0\ 0\ 1\ 1\ 1\ 1 & 79 \\ 0\ 0\ 1\ 1\ 1\ 1\ 1 & 31 \\ \hline 1\ 1\ 0\ 1\ 1\ 1\ 0_2 & \equiv 110_{10} \end{array}$$

Example 6.41 Subtract 01011_2 from 11001_2.
Solution:

$$\begin{array}{r} 1\ 1\ 0\ 0\ 1 \\ 0\ 1\ 0\ 1\ 1 \\ \hline 0\ 1\ 1\ 1\ 0_2 \end{array}$$

Example 6.42 Add 9 and 3 by means of signed binary nos.
Solution:

$$\begin{array}{rl} 0\ 0\ 0\ 0\ 1\ 0\ 0\ 1 & +9 \\ 0\ 0\ 0\ 0\ 0\ 0\ 1\ 1 & +3 \\ \hline 0\ 0\ 0\ 0\ 1\ 1\ 0\ 0 & +12 \end{array}$$

The first bit is 0 and hence the no is +ve as expected.

Example 6.43 Add –9 and +3 by means of signed binary nos. To represent –9, we take the 2's complements thus.
Solution:

$$\begin{array}{rr} 1\,1\,1\,1\,0\,1\,1\,1 & -9 \\ 0\,0\,0\,0\,0\,0\,1\,1 & +3 \\ \hline 1\,1\,1\,1\,1\,0\,1\,0 & -6 \end{array}$$

The sum is –ve since the Ist bit is 1. It is in 2's complement form and hence its magnitude is 0000110 which is 6 and hence the sum is –6.

Example 6.44 Add –9 and –3 by means of signed binary nos.
Solution:

$$\begin{array}{rr} 1\,1\,1\,1\,0\,1\,1\,1 & -9 \\ 1\,1\,1\,1\,1\,1\,0\,1 & -3 \\ \hline \text{Carry} \rightarrow 1\,1\,1\,1\,0\,1\,0\,0 & -12 \end{array}$$

In this solution, we discard the carry bit. The no is (–)ve and its magnitude in true binary form is 0001100 giving a decimal no. of –12.

Example 6.45 Add signed numbers 00001000, 00011111, 00001111 and 00101010.

Solution: For convenience the decimal equivalent numbers appear in the right hand column below.

$$\begin{array}{rlr} 0\,0\,0\,0\,1\,0\,0\,0 & & 8 \\ 0\,0\,0\,1\,1\,1\,1\,1 & \text{Ist sum} & 31 \\ \hline 0\,0\,1\,0\,0\,1\,1\,1 & & 39 \\ 0\,0\,0\,0\,1\,1\,1\,1 & & +15 \\ \hline 0\,0\,1\,1\,0\,1\,1\,0 & \text{2nd sum} & 54 \\ 0\,0\,1\,0\,1\,0\,1\,0 & & +42 \\ \hline 0\,1\,1\,0\,0\,0\,0\,0 & \text{3rd sum} & 96 \end{array}$$

Example 6.46 Subtract +9 from +12.
Solution:

$$\begin{array}{rll} 0\,0\,0\,0\,1\,1\,0\,0 & +12 & \\ 1\,1\,1\,1\,0\,1\,1\,1 & -9 & \text{(2's Complement)} \\ \hline 0\,0\,0\,0\,0\,0\,1\,1 & +3 & \end{array}$$

Discard the carry and we see that the number is positive with a magnitude of 3, *i.e.* the outcome is +3.

Example 6.47 Subtract +19 from −24.
Solution:

$$\begin{array}{rr} 1\,1\,1\,0\,1\,0\,0\,0 & -24 \\ 1\,1\,1\,0\,1\,1\,0\,1 & -19 \\ \hline 1\,1\,1\,0\,1\,0\,1\,0\,1 & -43 \end{array}$$

Discard the carry and we see that the number is (−)ve with a magnitude of 43, *i.e.* the outcome is − 43.

Example 6.48 Convert the actual numbers $(236)_8$ to decimal.
Solution:

$$\begin{aligned}(236)_8 &= 2 \times 8^2 + 3 \times 8^1 + 6 \times 8^0 \\ &= 2 \times 64 + 3 \times 8 + 6 \times 1 \\ &= 128 + 24 + 6 \\ &= (158)_{10}\end{aligned}$$

Each octal digit would require to be replaced by 3 binary digit.

Example 6.49 Convert $(125)_8$ to binary.
Solution: The binary for the Ist digit is 001
for the IInd digit is 010
for the IIIrd digit is 101
Hence, $(125)_8 = (001010101)_2$ in binary

Example 6.50 Convert decimal 6735 to binary.
Solution:

$$\begin{array}{r|rl} 2 & 6735 & \\ \hline 2 & 3367 & \to 1 \\ \hline 2 & 1683 & \to 1 \\ \hline 2 & 841 & \to 1 \\ \hline 2 & 420 & \to 1 \\ \hline 2 & 210 & \to 0 \\ \hline 2 & 105 & \to 0 \\ \hline 2 & 52 & \to 1 \\ \hline 2 & 26 & \to 0 \\ \hline 2 & 13 & \to 0 \\ \hline 2 & 6 & \to 1 \\ \hline 2 & 3 & \to 0 \\ \hline 2 & 1 & \to 1 \\ \hline & 0 & \to 1 \end{array}$$

$6735_{10} = (1101001001111)_2$

Example 6.51 An electrical control system uses 3 positional sensing devices, each of which produce 1 output when the position is confirmed. These devics are in to be used in conjuction with alogic network of NAND & OR gates and the output of network is to be 1 when two or more of the sensing devices are producing singals of I. Draw a network diagram of a suitable gate arrangement.

Solution: If we consider the possible combinations which satisfy the necessary conditions, it will be observed that there are four, *i.e.* any two devices or all three devices providing the appropriate signals, then.

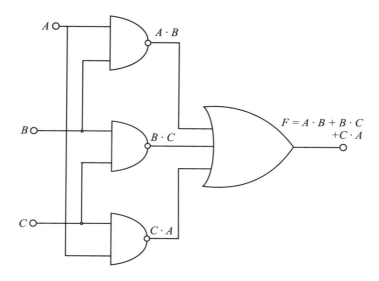

$$F = A \cdot B \cdot \overline{C} + A \cdot \overline{B} \cdot C + \overline{A} \cdot B \cdot C + A \cdot B \cdot C$$

The term $A \cdot B \cdot C$ can be repeated as of ten as desired, hence

$$F = A \cdot B \cdot \overline{C} + A \cdot B \cdot C + A \cdot \overline{B} \cdot C + A \cdot B \cdot C + \overline{A} \cdot B \cdot C + A \cdot B \cdot C$$

Using the IInd distributive rule

$$F = A \cdot B \cdot (\overline{C} + C) + A \cdot C (\overline{B} + B) + B \cdot C \cdot (\overline{A} + A)$$

but applying the identity $\quad A + \overline{A} = 1 B + \overline{B} = 1, C + \overline{C} = 1$

hence $\qquad F = A \cdot B + B \cdot C + C \cdot A$

The network which would effect this function is shown in above figure.

Example 6.52 Draw the ckt. of gates that could effect the function.

$$F = \overline{A \cdot B} + \overline{A \cdot C}$$

Simplify this function and hence redraw the ckt. that could effect it.

Solution: The gate ckt based on the original function is shown in this figure.

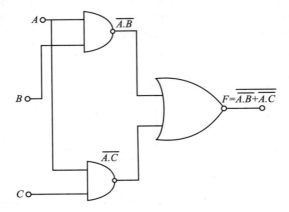

Using De Morgan's theorem

$$F = \overline{\overline{A \cdot B} + \overline{A \cdot C}}$$

$$= \overline{\overline{A \cdot B}} \cdot \overline{\overline{A \cdot C}}$$

$$= A \cdot B \cdot A \cdot C \text{ (Associative rule)}$$

$$= A \cdot B \cdot C$$

The simplified ckt. is shown in following figure.

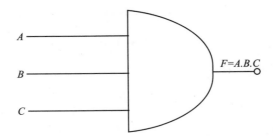

Example 6.53 Draw the ckt. of gates that would effect the function

$$F = \overline{A + B \cdot C}$$

Simplify this function and hence redraw the ckt. that could effect it.

Solution: The gate ckt. based on the original function is shown in above figure using De Morgan's theorem.

$$F = \overline{A + B \cdot C}$$
$$= \overline{A} \cdot \overline{(B \cdot C)}$$
$$= \overline{A} \cdot (\overline{B} + \overline{C})$$

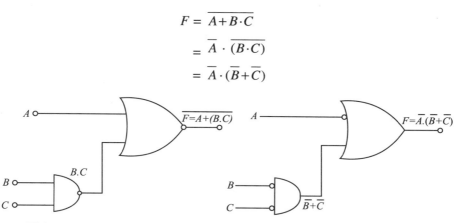

This can be realized by the network shown in above figure, which shows that, rather than there being a saving, we have involved the same number of gates with a greater number of invertices.

Example 6.54 Draw a logic in cicruit corporating any gates of your choice, which will produce an output 1 when its 2 and I/p's are different. Also draw a logic ckt. incorporating only NOR gates, which will perform the same function.

Solution: For such a requirement, the function takes the form

$$F = \overline{A} \cdot B + A \cdot \overline{B}$$

This is the NOT EQUIVALENT function and the logic ckt is shown in the figure given below.

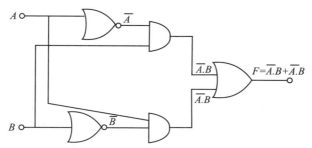

This can be converted directly into NOR logic gate ciruits, as shown in in the figure given below. Examination of the circuitry shows that 2 pairs of NOR gates and redundant since the output of each pair is the same as its input.

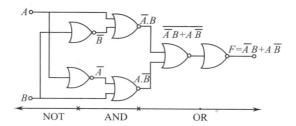

Example 6.55 Draw ckts which will generate the function.

$$F = B \cdot (\overline{A} + \overline{C}) + \overline{A} \cdot B$$

Using (a) = NOR gates
(b) = NAND gates

Solution: Given function $F = B \cdot (\overline{A} + \overline{C}) + \overline{A} \cdot B$

$$= B \cdot \overline{A} + B \cdot \overline{C} + \overline{A} \cdot B \quad \text{(Second distributive rule)}$$
$$= \overline{A} \cdot (B + B) + B\overline{C} \quad \text{(Second distributive rule)}$$
$$= \overline{A} + B \cdot \overline{C} \quad \text{(First rule of complementation)}$$

(a) For NOR gates complement of function is

$$F = \overline{\overline{A} + B \cdot \overline{C}}$$
$$= A \cdot \overline{(B \cdot \overline{C})} \quad \text{(De Morgan's theorem)}$$
$$= A \cdot (\overline{B} + C) \quad \text{(De Morgan's theorem)}$$
$$= A \cdot \overline{B} + A \cdot C \quad \text{(Second distributive rule)}$$

$A \cdot \overline{B}$ and $A \cdot C$ are generated separately giving the ckt. shown in the figure.

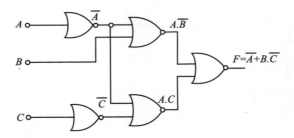

(b) For NAND gates

$$F = \overline{A} + B \cdot \overline{C}$$

I/p's to the final NAND gates are

$$\bar{\bar{A}} = A \quad \text{and} \quad \overline{\bar{B}\cdot\bar{C}} = B + C$$

$\bar{B} + C$ has to be generated separately, giving the ckt. shown in the figure.

$$F = \bar{A} + B \cdot \bar{C}$$

Example 6.56 Multiply the binary numbers 1101 and 0101.
Solution:

```
         1 1 0 1
         0 1 0 1
         ───────
         1 1 0 1
       0 0 0 0 ×
     1 1 0 1 × ×
   0 0 0 0 × × ×
   ─────────────
         1 1 1 1
   ─────────────
   1 0 0 0 0 0 1
```

Hence, $1101 \times 0101 = 1000001$

Example 6.57 Multiply the decimal numbers 27 and 10.
Solution:

```
         1 1 0 1 1      27
           1 0 1 0      10
         ─────────
         1 1 0 1 1
       1 1 0 1 1
       ───────────
       1   1 1
     1 0 0 0 0 1 1 1 0   = 270
```

Example 6.58 Multiply 45_{10} by 25_{10} using binary means.
Solution:

```
           1 0 1 1 0 1      45
           1 1 0 0 1        25
           ─────────
           1 0 1 1 0 1
       1 0 1 1 0 1
       ─────────────
       1 1 0 0 1 0 1 0 1
   1 0 1 1 0 1
   ─────────────────────
   1 0 0 0 1 1 0 0 1 0 1   ≡ 1125
```

Example 6.59 Divide 63_{10} by 9_{10} by means of binary numbers.
Solution:
$$63_{10} = 111111_2$$
$$9_{10} = 1001_2$$

```
                    1 1 1        Quotient
Divisor     1001 | 1 1 1 1 1 1   Dividend
                   1 0 0 1 ↓
                   ─────────
                   0 1 1 0 1
                   1 0 0 1
                   ─────────
                   0 1 0 0 1
                   1 0 0 1
                   ─────────
                   0 0 0 0       Remainder
```

Hence $\quad 111111 \div 1001 = 111$

and $\quad\quad\quad\ 63_{10} \div 9_{10} = 7_{10}$

In this case there was no remainder since 9 divides exactly into 63.

Example 6.60 Divide 61_{10} by 9_{10} by means of binary numbers.
Solution:
$$61_{10} = 111101_2$$
$$9_{10} = 1001_2$$

```
                    1 1 0        Quotient
Divisor     1001 | 1 1 1 1 0 1   Dividend
                   1 0 0 1 ↓
                   ─────────
                   0 1 1 0 0
                   1 0 0 1
                   ─────────
                   0 0 1 1 1     Remainder
```

Hence $\quad 111101 \div 1001 = 110$ remainder 111 and

$$61_{10} \div 9_{10} = 6_{10} \text{ remainder } 7_{10}$$

Example 6.61 Determine the 2's complement of 1101100_2 by applying the 1's complement.
Solution:
$$1111111 - 1101100 + 1 = 0010011 + 1$$
$$= 0010100$$

An alternative method of finding the 2's complement is as follows:

1. Start with the LSB and moving left, write down the bits as they appear upto and including the first 1.
2. Continuing left, write the 1's complement of the remaining bits.

Example 6.62 Determine the 2's complement of 1101100 directly. Split the no. to the left of the lowest 1.

Solution:

1101 change (invert) 0010

100 no change 100

Hence, 2's complement of 1101100 = 0010100.

Example 6.63 Determine the decimal value of the signed binary no. 10011010.

Solution: The weights are as follows

2^6	2^5	2^4	2^3	2^2	2^1	2^0
0	0	1	1	0	1	0

Summing the weights

$$16 + 8 + 2 = 26$$

The sign bit is 1, therefore the decimal no. is -26.

SUMMARY

1. **Number system:** Number systems are decimal, binary, octal and hexadecimal which have bases 10, 2, 8 and 16. The digits used are for decimal are 0 to 9, binary 0 and 1, octal 0 to 7, hexadecimal 0 to 9 and A to F. Binary digit is known as bit.

2. **Conversion of bases:**
    (*i*) Conversion from decimal to binary, decimal to octal and decimal to hexadecimal is done through division by the base number in which it is being converted to. The first remainder is Least Significant Digit and the last remainder is Last Significant Digit.
    (*ii*) Conversions from binary to decimal, octal to decimal and hexadecimal to decimal is obtained by a sum of various digits multiplied by their respective weights.
    (*iii*) Conversion from binary to octal and hexadecimal by making groups 3 bits and 4 bits respectively from right side of the binary number. Subsequently these groups are converted into the digits of respective bases.
    (*iv*) Conversion from octal and hexadecimal to binary is obtained by converting each digit to its 3-bit and 4-bit binary respectively.

(v) Conversion of hexadecimal to octal is done by converting each digit to its binary equivalent and binary equivalent is grouped with three bits each starting from LSB. The groups are converted into octal.

(vi) Fractional binary part conversion to fractional decimal part is obtained by a sum of various digits by the respective weights $-2^{-1}, 2^{-2}, 2^{-4}, ... , 2^{-n}$.

(vii) Fractional decimal number part conversion to fractional binary part is obtained by multiplying fractional decimal number by continued multiplication by 2 and noting down carry. First carry is MSB and the last carry is LSB.

3. **Binary Coded Decimal:** Each digit of decimal is converted into four binary bits and group seperation is maintained.

4. **Addition of Binary Numbers:** Digits are added from LHS *i.e.*, LSB and carry is taken to RHS for addition.

5. **Subtraction of Binary Numbers:** Digits are subtracted from LHS *i.e.*, LSB and digits are borrowed from RHS if needed.

6. **Logic Gates:** The basic logic gates are AND, OR and NAND and NOR logic gates are universal gates because any gate can construct with these gates. Logic gates operate in two conditions ON-OFF/TRUE-FALSE/HIGH-LOW/1-0 bistates.

7. **Logic Operators and Theorems:**

   Logic operators

OR	$F = A + B$
AND	$F = A \cdot B$
NOT	$F = \overline{A}$
Cummulative law	$A + B = B + A$
	$A \cdot B = B \cdot A$
Associative law	$A + (B + C) = (A + B) + C$
	$A \cdot (B \cdot C) = (A \cdot B) \cdot C$
Distributive law	$A + B \cdot C = (A + B) \cdot (A + C)$
	$A \cdot (B + C) = A \cdot B + A \cdot C$
De Morgan's law	$\overline{A \cdot B \cdot C} = \overline{A} + \overline{B} + \overline{C}$
	$\overline{A} \cdot \overline{B} \cdot \overline{C} = \overline{A + B + C}$

8. **Boolean Algebra:** It implements operations of a system of logic required for digital logics. Boolean algebraic statements use logic gates. These statements can take the form of algebaric equations, logic variables may have only two values 0 or 1.

9. Canonical Forms: If each term of a logic expression contains all variables, then it a called a canonical form. Canonical expression can be in sum of product (SOP) form, and each product term is known as minterm. Similarly, canonical expression can be in product of sums (POS) form, and each sum term is known as maxterm.

10. K-map: It is a graphical representation of fundamental products in a truth table. It requires number of squares in a rectangle or squares. Each square represents as minterm. A K-map has 2^n squares for n-variables.

11. Simplification of Boolean Expression Using K-map: Squares of showing 1 are combined to simplify and sum of product (SOP) form is used. In omplementary case squares showing 0 are combined to simplify and product of sums (POS) form is used.

EXERCISES

6.1. Write the difference in boolean algebra and ordinary algebra.
6.2. Write and prove:
   (*i*) Involution theorem.
   (*ii*) Absorption theorem.
6.3. How can you connect a NAND gate to make on inverter?
6.4. What are basic logic gates and what are universal logic gates?
6.5. Make the EX-OR gate with minimum number of NAND gates.
6.6. What are Boolean Postulates? State them.
6.7. Construct AND gate, NAND gate and OR gate with help of NOR gate.
6.8. What is the difference between canonical form and standard form? Also write the De Morgan's theorem.
6.9. What is a two inputs NAND gate called universal gate?
6.10. State and prove De Morgan's theoem.
6.11. Using 10's complement, subtract 72532 – 3250.   [**Ans.** 69282]
6.12. Use 2's complement and 1's complement to preform M–N where M = 1010100 and M = 1000100.
       [**Ans.** 0010000, 0010000]
6.13. Convert the following decimal numbers to binary:
   (*i*) 12.0625 (*ii*) 104 (*iii*) 673.23 (*iv*) 1998
       [**Ans.** (*i*) $(1100.0001)_2$, (*ii*) $(10011100010000)_2$,
       (*iii*) $(1110110010011101 0111)_2$, (*iv*) $(11111001110)_2$]
6.14. Convert the following binary numbers to decimal:
   (*i*) 10.10001 (*ii*) 101110.0101 (*iii*) 1101101.111
       [**Ans.** (*i*) (2.53125), (*ii*) (46.3125), (*iii*) (109.875)]

6.15. Convert decimal 255.225 to binary, octal and hexadecimal.

[**Ans.** $(11100001.0011100)_2$]

[**Ans.** $(341.16314)_8$ $(E1.39)_{16}$]

6.16. Converet 2AC5 . D to octal. [**Ans.** $(25305.64)_8$]

6.17. Obtain 1's and 2's complement of 1010101, 0111000.

[**Ans.** (i) (0101010, 1000111), (ii) (0101011, 1001000)]

6.18. Obtain 9's to 10's complement of 13579, 099001.

[**Ans.** (i) (45470, 49149), (ii) (86421, 90100)]

6.19. Find 10's complement of $(935)_{11}$. [**Ans.** (8873)]

6.20. Determine the value of X in the following equation:

$(11001)_2 = X_{10}$ [**Ans.** $X = 25$]

6.21. Convert hexadecimal B5A and 32F in to decimal no. [**Ans.** (i) 2906 (ii) 814]

6.22. Convert following binary numbers into hexadecimal nos.

(i) 11010110   (ii) 11111001 [**Ans.** (i) D_6 (ii) F_9]

6.23. Convert $(100101110 . 11101)_2$ to hexadecimal. [**Ans.** $(12E . E8)_{16}$]

6.24. Convert decimal no. 15, 65 into octal nos.

[**Ans.** (i) $15_{10} = 17_8$, (ii) $65_{10} = 101_8$]

6.25. Convert $(1001)_2$ to Gray code. [**Ans.** 11011]

6.26. Convert Gray code no (11010) into binary no. [**Ans.** 10011]

6.27. Convert decimal no. 245 to binary coded decimal (*BCD*).

[**Ans.** 001001000101]

6.28. Represent the decimal no. 8620 (*a*) in *BCD* (*b*) in X-3 in 8421 code (*d*) as a binary number.

[**Ans.** (*a*) $(1000011000100000)_{BCD}$, (*b*) $(1011100101010011)_{X-3}$,

(*c*) $(1110110000100000)_{8421}$, (*d*) $(10000110101100)_2$]

6.29. Obtain the weighted binary code for the basic digits using weights of 8421.

[**Ans.** 0 → 0000

1 → 0001

2 → 0010

3 → 0011

4 → 0100

5 → 0101

6 → 0110

7 → 0111

8 → 1011

9 → 1100

10 → 1101

11 → 1110]

6.30. Write $(13)_{10}$ in binary term and *BCD* term. [**Ans.** (i) $(13)_{10} = (1101)_2$,

(ii) $(00010011)_{BCD}$]

6.31. Assign a binary code in some orderly manne to the 52 playing cards. Use mini no. of bits.

[**Ans.** (2 bits for suit, 4 bits for no. J = 1011, Q = 1100 and K = 1101)]

6.32. Write the logic equation for the output P in terms of I/p's A, B, C and D.

[**Ans.** $P = (A + B) \cdot C + D$]

6.33. What is the logic state of Y (see. in fig.) when A, B, C, D are

(*i*) 0000 (*ii*) 0110 (*iii*) 1011 (*iv*) 1110 (*v*) 1111   [**Ans.** 0, 0, 1, 1, 0]

6.34. Add the following binary nos.

(*i*) 101010 and 110110 (*ii*) 110110 and 111100   [**Ans.** 110000, 1110010]

6.35. Express the following functions in a sum of minterm and product of maxterms.

(*i*) $F(x, y, z) = 1$ (*ii*) $(A, B, C, D) = D \cdot (\overline{A} + B) + \overline{B} \cdot D$.

[**Ans.** (*i*) There is no maxterm in it,

(*ii*) Max. terms = $\pi(0, 2, 4, 6, 8, 10, 12, 4)$]

6.36. Convert the following others canonical terms.

(*a*) $f(x, y, z) = \Sigma(1, 3, 7)$

(*b*) $f(x, y, z) = \pi(0, 3, 6, 7)$

[**Ans.** (*a*) = $\pi(0, 2, 4, 5, 6)$, (*b*) = $\Sigma(1, 2, 4, 5)$]

6.37. Obtain the simplified expression in sum of products for the following Boolean function using K-map.

(*i*) $f = \overline{x}yz + x\overline{yz} + xyz + z\overline{yz}$

(*ii*) $f(x, y, z) = \Sigma m(0, 2, 4, 5, 6)$

(*iii*) $f(w, x, y, z) = \Sigma m(91, 2, 4, 5, 6, 8, 9, 12, 13, 14)$

[**Ans.** (*i*) $f = x\overline{z} + yz$, (*ii*) $f = \overline{z} + x\overline{y}$, (*iii*) $f = \overline{y} + \overline{w}\overline{z} + x\overline{z}$]

6.38. For the following truth table:

x	y	z	f_1	f_2
0	0	0	0	0
0	0	1	1	0
0	1	0	1	0
0	1	1	0	1
1	0	0	1	0
1	0	1	0	1
1	1	0	0	1
1	1	1	1	1

(a) Express f_1 and f_2 in product of max.term.

(b) Obtain the simplified function in SOP form using K-map.

(c) Obtain the simplification function in POS form using K-map.

[**Ans.** (a) $f_1 = M_0 \cdot M_3 \cdot M_5 \cdot M_6$, $f_2 = M_0 \cdot M_1 \cdot M_2 \cdot M_4$,

(b) SOP = $yz + xy + \overline{xy}z$, (c) POS = $(x + z) \cdot (x + y) \cdot (y + z)$]

6.39. Simplify each of the following boolean function (+) using the donot care condition (d) in (i) sum of products and (ii) product of sums.

[**Ans.** (i) SOP $F = \overline{A}C + \overline{B}\overline{D}$ POS $F = (\overline{B} + C) \cdot \overline{A}(C + \overline{D})$,

(ii) SOP $F = \overline{XZ} + \overline{W}Z$ POS $\overline{F} = (\overline{X} + Z)(\overline{W} + Z)$]

7

ELECTRONICS INSTRUMENTS

Analog instruments are rapidly being replaced by digital instruments. Measurement of voltage, current, resistanced phase and frequency are the paramenters of interest. Digital voltmeter, digital multimeter will be considered and Cathode Ray Oscilloscope (**CRO**) will be considered.

A digital instrument building block is shown in Fig. 7.1 Analog signal is converted into digital signal for being measured by digital technique.

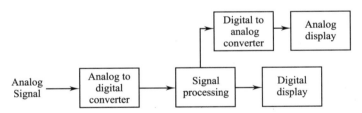

Fig. 7.1 Digital Instrument Building Block.

The display block may be analog or digital in nature. If an analog display is needed, a digital to analong converter will be used.

7.1 DIGITAL VOLTMETERS (DVMs)

Digital voltmeters convert analog voltage signals into a digital output signal. This digital output signal is displayed on the front panel. Thus, **DVMs** have speed of measurement, accuracy, automation and programming feasibility. It may be noted that analog voltmeters have pointers and continuous scale which are prone to human errors and parallax errors. But, DVMs are free from such errors. DVMs are small in size, power requirement and have reduced prices due to development of IC's. There are several techniques of converting analog to digital signal, therefore, DVMs are classified based on these methods. Ramp-type DVMs and **staircase-ramp** DVMs will be considered here.

7.1.1 Ramp-type DVMs

Linear ramp technique measures time taken to rise zero volt to the level of the input voltage or to decrease from the level of voltage to zero volt. An electronic time-interval counter is used to measure the time interval. The time interval is displayed as number of digits on a display. Fig. 7.2 shows voltage to time conversion. A negative ramp voltage is initiated at the start of the measurement. This ramp voltage is continuously checked for first and second coincidences. First coincidence is equal to the voltage to be measured and the second is equal to zero. The time interval between the first and second coincidence is measured by counting the clock pulses by electronic counter. This pulse count is the direct measure of the input voltage.

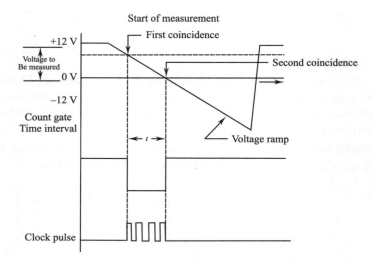

Fig. 7.2 Voltage-to-time Conversion.

Figure 7.2 shows the block diagram of a ramp-type DVM. The dc voltage to be measured is input to the Ranging and Attenuator which output different rangers of measurement. The rate of measurement cycles are initiated by the "Sample -rate multivibrator (MV)". A control on the front panel adjusts the oscillator of MV for few cycles per second to as high as 1000 or ever more.

A initiating pulse to the "Ramp generator" for starting the ramp-voltage, is provided by the "Sample-rate MV". The ramp voltage is fed to both the "Input comparator" and "Ground Comparator". At the same moment, the sample a ate "MV" gives reset pulse to the Decode Counting units (DCVs) making them to their state and this removes momentarily "Digital display".

There is continuous comparator between the input voltage and the ramp voltage by the "Input Comparator". When the ramp voltage equals the voltage to be measured, then the comparator generate "Start pulse" and this opens the "Gate". The "Clock-pulse generator" generated pulse are allowed to go by the "Gate" to the "Decode counting units (DCVs). The "Ground comparator" generate a "Stop pulse" when the continuous reduction of the ramp voltage with respect to time with ground potential at 0 V. This closes the "Gate". The number of pulses passed through the "Gate" are totalised by the "DCV's". The measured input voltage can be seen on the "Digital display" unit.

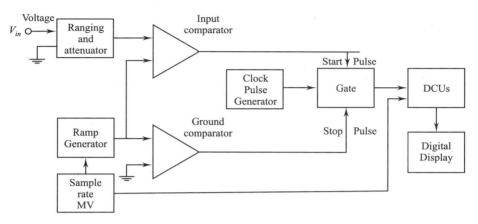

Fig. 7.2a Block diagram of a ramp-type DVM.

7.1.2 Staircase-ramp DVMs

In this case, the voltage to be measured (V_{in}) is compared with an internally generated "Staircase-ramp voltage". V_{in} is converted to a BCD-Code representation which is subsequently decoded and displayed on a display of digital type.

As an example a staircase-ramp DVM is shown is Figure 7.3. It can be seen that the block diagram has four digit which can be increased if needed.

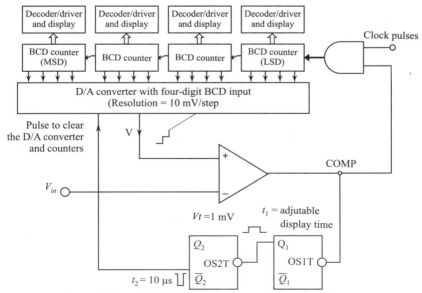

Fig. 7.3 Four-digit Staircase-ramp type Digital Voltmeter.

Consider that for every step of digital input from BCD counter, the D/A converter produces a step of 10mV. Thus, the counters run up from 0000 to 9999 and the staircase voltage (V) rises from 0 mV to 99990 mV i.e., 99.99 V which is the maximum input voltage of the DVM.

A 4.5 kHz relaxation oscillator generates. The clock pulses which are gated through an AND gate into the counters. COMP (comparator output) signal enables the AND gate. LSD counter gives a carry pulse to the lens decode counter at every tenth input pulse.

The lens counter gives its own carry to the hundreds counter, and so on. The display is held for time duration t_1 for observation after the counting ends. As long as the input V_{in} > staircase-ramp voltage V, the comparator output is 1 and the AND gate is open for the clock pulses to pass the counters, the counter advances a step and V goes up another 10mV with each clock pulse. When $V > V_{in}$ by a value of 10mV, COMP goes to 0 which disables the AND gate.

When COMP goes 0, it also triggers one shot device $OS1$ whose output Q_1 becomes 1 and it effectively hold the display for time duration t_1. When Q_1 input to $OS2$ goes 0, it is triggered and causes Q_2 to clear the A/D converter and BD counters to the 0 state. The pulse is of duration (t_2) about 10 ms. After the clearance of D/A converter, $V_{in} > V$ and whole counting process is repeated. Each reading is displayed until new reading is completed.

7.2 DIGITAL MULTIMETERS (DMMS)

A digital multimeter is used for measurement of ac/dc voltage, ac/dc currents and resistances. The basic circuit of a DMM is shown in Figure 7.4.

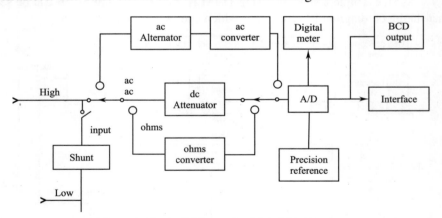

Fig. 7.4 Block diagram of a Digital Multimeter.

The basic circuit is a dc digital voltmeter. Current is converted to voltage. DC current is passed through a precision law shunt resistance while ac current is converted into dc by using rectifiers and filters. In the case of resistance measurement, DMM includes a precision low current source which is applied across the resistance to be measured. The dc voltage so developed is digitised and displayed

Electronics Instruments 273

as ohms. A typical DMM is shown in Fig. 7.5. It has power switch, function switch, Hz/% select button, data hold button, relative button, select button, Range hold button, Display, test leads *etc*. Test leads have two test probes red and black, finger guards and test pins, *etc*.

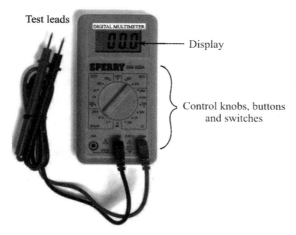

Fig. 7.5 A typical digital multimeter.

7.3 CATHODE RAY OSCILLOSCOPE (CRO)

The CRO allows the amplitude of electrical signals, whether they are voltage current, or power to be displayed as a function of time. A block diagram of a general purpose CRO is shown in Fig. 7.6.

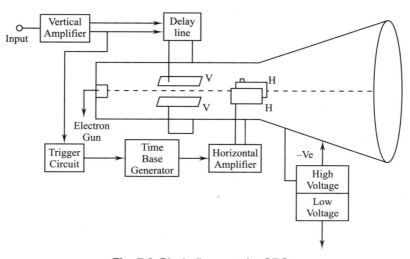

Fig. 7.6 Block diagram of a CRO.

CRO is comprised of **CRT**, Vertical amplifier, Delay line, Time base, Horizontal amplifier, Trigger circuit and Power supply units.

Cathode Ray Tube (CRT) is basically an electron beam voltmeter. The electron gun generates a narrow eletron beam which is bombarded on the screen of the tube. The screen is the external flat end of the glass tube which is chemically treated to form a flurescent screen. The screen glows at the point of collision *i.e.*, produces a bright spot. The electron beam is deflected at a constant rate relative of time along the *x*-axis and is deflected along the *y*-axis in response to an stimulus such as voltage. This produces a time-dependent variation of the input voltage. As the electron has practically no weight *i.e.* no inertia, hence, the beam of electrons can be moved to follow waveforms varying at a rate of millions of times/second. Thus, electron beam faithfully follows rapid variations in signal voltage and trans a visible path on the CRT screen. In this way, rapid variations, pulsations or transients are reproduced and the operator can observe the waveform as well as measure amplitude at any instant of time.

Vertical amplifier is a wide band amplifier used to amplify signals in the vertical section. Delay time is used to delay the signal for sometime in the vertical sections. Time base is used to generate the santooth voltage required to deflect the beam in this horizontal section. Horizontal amplifier is used to simplify the santooth voltage before it is applied to horizontal deflection plants. Trigger circuit is used to convert the incoming signal into trigger pulses so that the input signal and sweep frequency can be eynchronised. There are two power supplies, a –ve High voltage (HV) supply and a +ve Low Voltage (LV) supply. These two voltages are generated in the CRO. The +ve voltage supply is from +300 to 400 V. The –ve high voltage supply is from –1000 to –1500V. This voltage is passed through a bleeder resistor for intensity, focus and positioning controls.

A front control panel of CRO is shown in Fig. 7.7. The boards and switches are identified for each function of CRO.

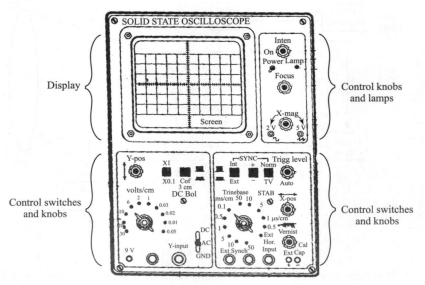

Fig. 7.7 A Front Control panel an CRO.

7.4 MEASUREMENTS USING CRO

7.4.1 Measurement of Voltage

A dc voltage to be measured is given to vertical defection plates. The displacement of the spot on the screen is measured. The displacement multiplied by the deflection sensitivity which gives the magnitude of dc voltage.

An ac voltage to be measured is also given to the vertical deflection plates. The length of the straight line trace obtained on the screen is measured. This length is multiplied with deflection sensitivity in V/cm giving the peak-to-peak value of the ac voltage. This value is divided by $2\sqrt{2}$ giving the rms value of the ac voltage to be measured.

7.4.2 Measurement of Current

Current is measured indirectly with the help of CRO. The current to be measured is passed through a suitable known resistor. The voltage developed across the resistor is measured using CRO. The voltage is divided by the resistance value gives the current value.

7.4.3 Measurement of Phase Difference

CRO can determine phase difference between two sine - waves of the same frequency. The two voltages are simultaneously applied to the two sets of deflection plates. This gives the Lissajous pattern on the screen as an ellipse as shown in Fig. 7.8.

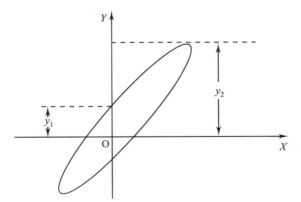

Fig. 7.8 Ellipse created in measurement of phase difference between two sine-wave.

The Fig. is centred by adjusting the x-shift and y-shift controls. The intercepts y_1 and y_2 are measured. The phase difference is given by:

Phase difference, $\theta = \sin^{-1}\left(\dfrac{y_1}{y_2}\right)$.

7.4.4 Measurement of Frequency

The ac voltage is displayed on the CRO and measurement of its time period is done using the calibrated time base. The frequency is calculated as the inverse of this time period.

Time period of the ac waveform, T = number of dividions in one cycle $\left(\dfrac{\text{Time}}{\text{division}}\right)$ where, T = time period of the ac save in sec. frequency, $f = \dfrac{1}{T}$ Hz.

SUMMARY

1. The building blocks of any digital instruments are A/D Converter, D/A Converter, Single processing unit, Analog display, Digital display units etc.
2. Digital Voltmeters (DVMS) convert analog voltage signals into a digital output signal which is displayed. There are various types of DVM's and the important ones are amp. type and staircase-ramp type.
3. Digital multimeters use electronic circuits, such as instrumentation amplifier to amplify the voltage to be measured.
4. Digital multimeters are used to measure ac/dc voltages, ac/dc currents and resistances. The building blocks are ac/dc alternators, A/D converter, Digital Display *etc.*
5. Cathode Ray Oscilloscope (CRO) is a very fast x - y plotter and is used to display voltage wave forms. The "stylus" of the "plotter" is a luminous spot which moves on the screen in response to the input voltages.
6. CRO is comprised of CRU, vertical amplifier, delay line, time box, horizontal amplifier, trigger circuit about and power supply units.
7. Lissajous pattern is formed on the screen of a CRO, when two sine wave voltages are simultaneously applied to the two sets of deflection plates. The phase difference between the sine waves applied is determined from this pattern.

EXERCISES

7.1 What are the advantages of digital instruments over analog instruments?

7.2 Explain working principle of digital voltmeter with block diagrams.

7.3 What are the applications of a digital multimeter? Explain DMM working principle with block diagram.

7.4 What are the uses of CROS? How is phase difference between two sine waves are measured using a CRO.

7.5 Describe basic building blocks of a CRO and its working principle.

7.6 A sinusoidal voltage is applied to Y-input of a CRO. Its vertical amplifier sensitivity is set at 1 V/cm. A straight line trace of length 6.2 cm is obtained on the screen. What is the rms value of the sinusoidal voltage?

Appendix-A

SYMBOLS, ABBREVIATIONS AND DIAGRAMMATIC SYMBOLS

Table A-1 Abbreviations for Multiples and Sub-Multiples

Symbol	Abbreviations	Multiples
T	tera	10^{12}
G	giga	10^{9}
M	Mega or meg	10^{6}
k	kilo	10^{3}
d	deci	10^{-1}
c	centi	10^{-2}
m	milli	10^{-3}
µ	micro	10^{-6}
n	nano	10^{-9}
p	pico	10^{-12}

Table A-2 Miscellaneous Symbols

Term	Symbol
Approximately equal to	$\simeq$
Proportional to	$\propto$
Infinity	∞
Sum of	Σ
Increment of finite difference operator	Δ, δ
Greater than	$<$
Less than	$>$
Much greater than	$\gg$
Much less than	$\ll$

Contd...

Term	Symbol
Base of natural logarithms	e
Common logarithm of x	log x
Natural logarithm of x	ln x
Complex operator (-1)	j
Temperature	θ
Time constant	T
Efficiency	η
Per unit	p.u.

Table A-3 Greek Alphabet

Term	Capital	Lowercase
alpha	A	α
beta	B	β
gamma	Γ	γ
delta	Δ	δ
epsilon	E	ε
zeta	Z	ζ
eta	H	
theta	Θ	θ
iota	I	ι
kappa	K	κ
lambda	Λ	λ
mu	M	μ
nu	N	ν
xi	Ξ	ξ
omicron	O	o
pi	Π	π
rho	P	ρ
sigma	Σ	σ
tau	T	τ
upsilon	Y	υ
phi	Φ	ϕ
chi	X	χ
psi	Ψ	ψ
omega	Ω	ω

Table A-4 Electrical Units

Quantity	Quantity symbol	Unit	Unit symbol
Admittance	Y	siemens	S
Angular velocity	ω	radian per second	rad/s
Capacitance	C	farad	F
		microfarad	μF
		picofarad	pF
Charge on Quantity of Electricity	Q	coulomb	C
Conductance	G	siemens	S
Conductivity	σ	siemens per metre	S/m
Current			
Steady or r.m.s. value	I	ampere	A
		milliampere	mA
		microampere	μA
Instantaneous value	i		
Maximum value	I_m		
Current density	£	ampere per square metre	A/m^2
Difference of potential			
steady or r.m.s. value	V	volt	V
		millivolt	mV
		kilovolt	kV
Instantaneous value	v		
Maximum value	V_m		
Electric field strength	E	volt per metre	V/m
Electric flux	Ψ	coulomb	C
Electric flux density	D	coulomb per square metre	C/m^2
Electromotive force			
Steady or r.m.s. value	E	volt	V
Instantaneous value	ε		
Maximum value	E_m		
Energy	W	joule	j
		kilojoule	kJ
		megajoule	MJ
		watt hour	W h
		kilowatt hour	kwh
		electronvolt	eV
Force	F	newton	N
Frequency	f	hertz	Hz
		kilohertz	kHz
		megahertz	MHz

Contd...

Quantity	Quantity symbol	Unit	Unit symbol
Impedance	Z	ohm	Ω
Inductance, self	L	henry (plural, henrys)	H
Inductance, mutual	M	henry (plural, henrys)	H
Magnetic field strength	H	ampere per metre	A/m
		ampere turns per metre	At.m
Magnetic flux	Φ	weber	Wb
Magnetic flux density	B	tesla	T

Table A-5 Magnetic Units

Quantity	Quantity symbol	Unit	Unit symbol
Magnetic flux linkage	ψ	Weber	Wb
Magnetomotive force	F	ampere	A
		ampere turns	At
permeability of free space or Magnetic constant	μ_0	henry per metre	H/m
Permeability, relative	μ_r		
Permeability, absolute	μ		
Permittivity of free space of Electric constant	ε_0	farad per meter	F/m
Permittivity, relative	ε_r		
Permittivity, absolute	ε		
Power	P	Watt	W
		kilowatt	kW
		megawatt	MW
Power, apparent	S	voltampere	VA
Power, reactive	Q	var	var
Reactance	X	ohm	Ω
Reactive voltampere	Q		
Reluctance	S	ampere per weber	A/Wb
Resistance	R	ohm	Ω
		microhm	μΩ
		megaohm	MΩ
Resistivity	ρ	ohm metre	Ωm
Speed, linear	u	metres per second	m/s
Speed, rotaional	ω	radians per second	rad/s
	n	revolutions per second	r/s
	N	revolutions per minutes	r/min
		microhm metre	μΩ m
Susceptance	B	siemens	S

Contd...

Quantity	Quantity symbol	Unit	Unit symbol
Torque	T	Newton metre	Nm
Voltampere	-	voltampere	VA
		kilovoltampere	kVA
Wavelength	λ	metre	m
		micrometre	µm

Table A-6 Light Units

Quantity	Quantity symbol	Unit	Unit symbol
Illuminance	E	lux	lx
Luminance (objective brightness)	L	candela per square metre	cd/m^2
Luminous flux	Φ	lumen	lm
Luminous intensity	I	candela	cd
luminous efficacy	-	lumen per watt	lm/W

Table A-7 Section of graphical symbols from BS 3939

Description	Symbol	Description	Symbol
Direct current or steady voltage	—	Alternating	∼
Positive polarity	+	Negative polarity	—
Primary or secondary cell	—⊣⊢—	Battery of primary or secodary cells	—⊣|||⊢—
Fixed resistor	—▭—	Variable resistor	
		Resistor with moving contact	
Filament lamp		Crossing of conductor symbols on a diagram (no electrical connection)	+
Junction of conductors			
Double junction of conductors		Earth	⏚
Capacitor general symbol	⊥T	Polarized capacitor	
Winding	⌒⌒⌒	Inductor and core	
Transformer		Ammeter	Ⓐ
Voltmeter	Ⓥ	Wattmeter	Ⓦ
Galvanometer		Motor	Ⓜ

Contd...

Description	Symbol	Description	Symbol
Generator		Make contact (Normally open)	
Break contact (normally closed)		Rectifier	
Zener diode		P-N-P transistor	
N-P-N transistor		N-channel JUGFET	
P-channel JUGFET		N-channel IGFET	
P-channel IGFET		Amplifier	
Thyristor		MOSFET	
IGBT		GTO	
Binary logic units			
AND		OR	
NOT		NAND	
NOR			

Appendix-B

UNITS AND CONVERSION FACTORS

Table B-1

Quantity	symbol	Unit of Unit	Dimension
Fundamental			
Length	l, L	meter	L
Mass	m, M	kilogram	M
Time	t	second	T
Current	i, I	ampere	I
Mechanical			
Force	F	newton	MLT^{-3}
Torque	T	newton-meter	ML^2T^{-2}
Angular displacement	θ	radian	—
Velocity	v	meter/second	LT^{-1}
Angular velocity	ω	radian/second	T^{-1}
Acceleration	a	meter/second^2	LT^{-2}
Angular acceleration	α	radian/second^2	T^{-2}
Spring constant (translation)	K	newton/meter	MT^{-2}
Spring constant (rotational)	K	newton/meter	ML^2T^{-2}
Damping coefficient (translational)	D, F	newton-second/meter	MT^{-1}
Damping coefficient (rotational)	D, F	newton-second/meter	ML^2T^{-1}
Moment of inertia	J	kilogram-meter^2	ML^2
Energy	W	Joule (watt-second)	ML^2T^{-2}
Power	P	Watt	ML^2T^{-3}

Contd...

Quantity	symbol	Unit of Unit	Dimension
Electrical			
Charge	q, Q	coulomb	TI
Electric potential	v, V, E	volt	$ML^2T^{-3}I^{-1}$
Electric field intensity	f	volt/meter (or newton/coulomb)	$MLT^{-3}I^{-1}$
Electric flux density	D	coulomb/meter^2	$L^{-2}TI$
Electric flux	ψ, Q	coulomb	TI
Resistance	R	ohm	$ML^2T^{-3}I^{-2}$
Resistivity	ρ	ohm-meter	$ML^3T^{-3}I^{-2}$
Capacitance	C	farad	$M^{-1}L^{-2}T^4I^2$
Permittivity	ε	farad/meter	$M^{-1}L^{-3}T^4I^2$
Susceptance	S	siemens (ampere/volt)	$M^{-1}L^{-2}T^3I^2$
Magnetic			
Magnetomotive force	$£$	ampere(–turn)	I
Magnetic field intensity	H	ampere(–turn)/meter	$L^{-1}I$
Magnetic flux	ϕ	weber	$ML^2T^{-2}I^{-1}$
Magnetic flux density	B	tesla	$MT^{-2}I^{-1}$
Magnetic flux linkages	λ	weber-turn	$ML^1T^{-1}I^{-1}$
Inductance	L	henry	$ML^2T^{-2}I^{-2}$
Permeability	μ	henry/meter	$MLT^{-2}I^{-2}$
Reluctance	R	ampere/weber	$M^{-1}L^{-2}T^2I^4$

Table B-2 Conversion Factors

Quantity	Multiply number of :	By:	To obtain
Length	meters	100	centimeters
	meters	39.37	inches
	meters	3.281	feets
	inches	0.0254	meters
	inches	2.54	centimeters
	feet	0.3048	meters
Force	newtons	0.2248	pounds
	newtons	10^5	dynes
	pounds	4.45	newtons
	pounds	4.45×10^5	dynes
	dynes	10^{-5}	newtons
	dynes	2.248×10^{-5}	pounds
Torque	newton-meters	0.7376	pound-feet
	newton-meters	10^7	dyne-centimeters
	pound-feet	1.356	newton-meters
	dyne-centimeters	10^{-7}	newton-meters

Contd...

Appendix

Quantity	Multiply number of :	By:	To obtain
Energy	joules (watt-seconds)	0.7376	foot-pounds
	joules	2.778×10^{-7}	kilowatt-hours
	joules	10^7	ergs
	joules	9.480×10^{-4}	British thermal units
	foot-pounds	1.356	joules
	electron-volts	1.6×10^{-19}	joules
Power	watts	0.7376	foot-pounds/second
	watts	1.341×10^{-3}	horsepower
	horsepower	745.7	watts
	horsepower	0.7457	kilowatts
	foot-pounds/second	1.356	watts

Appendix-C

PERIODIC TABLE OF THE ELEMENTS

	I	II	III	IV	V	VI	VII	VIII	
1.	H 1 1.0081								He 2 4.002
2.	Li 3 6.940	Be 4 9.02	B 5 10.82	C 6 12.01	N 7 14.008	O 8 16.000	F 9 19.00		Ne 10 20.183
3.	Na 11 22.997	Mg 12 24.32	Al 13 26.97	Si 14 28.06	P 15 31.02	S 16 32.06	Cl 17 35.457		Ar 18 39.994
4.	K 19 39.096	Ca 20 40.08	Sc 21 45.10	Ti 22 47.90	V 23 50.95	Cr 24 52.01	Mn 25 54.93	Fe 26 Co 27 55.84 58.94	Ni 28 58.69
	Ca 29 63.57	Zn 30 65.38	Ga 31 69.72	Ge 32 72.6	As 33 74.91	Se 34 78.96	Br 35 79.916		Kr 36 83.7
5.	Rb 37 85.48	Sr 38 87.63	Y 39 88.92	Zr 40 91.22	Cb 41 92.91	Mo 42 96.0	Te 43	Ru 44 Rh 45 101.7 102.91	Pd 46 106.7
	Ag 47 107.880	Cd 48 112.41	In 49 114.76	Sn 50 118.70	Sb 51 121.76	Te 52 127.61	I 53 126.92		Xe 54 131.3
6.	Ce 55 132.91	Ba 56 137.36	La 57 138.92	Hf 72 178.6	Ta 73 180.88	W 74 184.0	Re 75 186.31	Os 76 Ir 77 191.5 193.1	Pt 78 195.23
	Au 79 197.2	Hg 80 200.61	Ti 81 204.39	Pb 82 207.21	Bi 83 209.00	Po 84 -	At 85 -		Rn 86 222
7.	Fr 87 -	Ra 88 226.05	Ac 89 -	Th 90 232.12	Pa 91 231	U 92 238.07			

Note: The number to the right of the symbol for the element gives the atomic number. The number below the symbol for the element gives the atomic weight. This table does not include the rare earths and the synthetically produced elements above 92.

Appendix-D

CONDUCTION PROPERTIES OF COMMON METALS

Table D-1 Resistivity and resistance temperature coefficient

Material	Resistivity, ρ microhm-cm at 20°C	Resistivity, ρ ohm-cir. mils per foot at 20°C	Resistance temp. coefficient at 20°C, α
Aluminum	2.828	-	0.0039
Brass	-	40	0.0017
Copper (std. annealed)	1.724	10.37	0.00393
Nichrome	100	-	0.0004
Silver	1.63	-	0.0038
Tungston	-	33.2	0.0045

Table D-2 Round Copper-Wire DATA

AWG number	Area (cir.mils)	Resistance (Ω/1000 ft)	Weight (ib/100 ft)	Allowable Current* (A)
0000	212,000	0.0490	640	358
000	168,000	0.0618	508	310
00	133,000	0.0779	402	267
0	106,000	0.0983	319	230
1	83,700	0.1240	253	196
2	66,400	0.156	201	170
3	52,600	0.197	159	146

Contd...

AWG number	Area (cir.mils)	Resistance (Ω/1000 ft)	Weight (ib/100 ft)	Allowable Current* (A)
4	41,700	0.248	126	125
5	33,100	0.313	100	110
6	26,300	0.395	79.5	94
8	16,500	0.628	50	69
10	10,400	0.999	31.4	50
12	6,530	1.59	19.8	37
14	4,110	2.52	12.4	29

* For type RH insulation—National Electrical Code*.

Table D-3 Standard Values of Commercially Available Resistors

Ohms (Ω)				Kilohms ($k\Omega$)			Megaohms ($M\Omega$)	
0.10	1.0	10	100	1000	10	100	1.0	10.0
0.11	1.1	11	110	1100	11	110	1.1	11.0
0.12	1.2	12	120	1200	12	120	1.2	12.0
0.13	1.3	13	130	1300	13	130	1.3	13.0
0.15	1.5	15	150	1500	15	150	1.5	15.0
0.16	1.6	16	160	1600	16	160	1.6	16.0
0.18	1.8	18	180	1800	18	180	1.8	18.0
0.20	2.0	20	200	2000	20	200	2.0	20.0
0.22	2.2	22	220	2200	22	220	2.2	22.0
0.24	2.4	24	240	2400	24	240	2.4	
0.27	2.7	27	270	2700	27	270	2.7	
0.30	3.0	30	300	3000	30	300	3.0	
0.33	3.3	33	330	3300	33	330	3.3	
0.36	3.6	36	360	3600	36	360	3.6	
0.39	3.9	39	390	3900	39	390	3.9	
0.43	4.3	43	430	4300	43	430	4.3	
0.47	4.7	47	470	4700	47	470	4.7	
0.51	5.1	51	510	5100	51	510	5.1	
0.56	5.6	56	560	5600	56	560	5.6	
0.62	6.2	62	620	6200	62	620	6.2	
0.68	6.8	68	680	6800	68	680	6.8	
0.75	7.5	75	750	7500	75	750	7.5	
0.82	8.2	82	820	8200	82	820	8.2	
0.91	9.1	91	910	9100	91	910	9.1	

Table D-4 Typical Capacitor Values

pF						μF		
10	100	1000	10,000	0.10	1.0	10	100	1000
12	120	1200						
15	150	1500	15000	0.15	1.5	18	180	1800
22	220	2200	20,000	0.22	2.2	22	220	2200
27	270	2700						
33	330	3300	33,000	0.33	3.3	33	330	3300
39	390	3900						
47	470	4700	47,000	0.47	4.7	47	470	4700
56	560	5600						
68	680	6800	68,000	0.68	6.8			
82	820	8200						

Appendix-E

RIPPLE FACTOR AND VOLTAGE CALCULATION

E.1 RIPPLE FACTOR OF RECTIFIER

The ripple factor of a voltage is defined by

$$r = \frac{\text{rms value of ac component of signal}}{\text{average value of signal}}$$

which can be expressed as

$$r = \frac{V_r(rms)}{V_{dc}}$$

Since the ac voltage component of a signal containing a dc level is

$$v_{ac} = v - V_{dc}$$

the rms value of the ac component is

$$V_r(rms) = \left[\frac{1}{2\pi}\int_0^{2\pi} v_{ac}^2 \, d\theta\right]^{1/2}$$

$$= \left[\frac{1}{2\pi}\int_0^{2\pi} (v - V_{dc})^2 \, d\theta\right]^{1/2}$$

$$= \left[\frac{1}{2\pi}\int_0^{2\pi} (v^2 - 2vV_{dc} + V_{dc}^2) \, d\theta\right]^{1/2}$$

$$= [V^2(rms) - 2V_{dc}^2 + V_{dc}^2]^{1/2}$$

$$= [V^2(rms) - V_{dc}^2]^{1/2}$$

where $V(rms)$ is the rms value of the total voltage. For the half-wave rectified signal,

$$V_r(rms) = [V^2(rms) - V^2_{dc}]^{1/2}$$

$$= \left[\left(\frac{V_m}{2}\right)^2 - \left(\frac{V_m}{\pi}\right)^2\right]^{1/2}$$

$$= V_m\left[\left(\frac{1}{2}\right)^2 - \left(\frac{1}{\pi}\right)^2\right]^{1/2}$$

$$\boxed{V_r(rms) = 0.385\,V_m \text{ (half-wave)}} \qquad \text{E.1}$$

For the full-wave rectified signal,

$$V_r(rms) = [V^2(rms) - V^2_{dc}]^{1/2}$$

$$= \left[\left(\frac{V_m}{\sqrt{2}}\right)^2 - \left(\frac{2V_m}{\pi}\right)^2\right]^{1/2}$$

$$= V_m\left(\frac{1}{2} - \frac{4}{\pi^2}\right)^{1/2}$$

$$\boxed{V_r(rms) = 0.308\,V_m \quad \text{(Full Wave)}} \qquad \text{E.2}$$

E.2 RIPPLE VOLTAGE OF CAPACITOR FILTER

Assuming a triangular ripple waveform approximation as shown in Fig. E.1, We can write (*see Fig.* E.2)

$$V_{dc} = V_m - \frac{V_r(p-p)}{2} \qquad \text{E.3}$$

During capacitor discharge, the voltage change across C is

$$V_r(p-p) = \frac{I_{dc}T_2}{C} \qquad \text{E.4}$$

From the triangular waveform in Fig. E.1

$$V_r(rms) = \frac{V_r(p-p)}{2\sqrt{3}} \qquad \text{E.5}$$

(Obtained by calculations not shown)

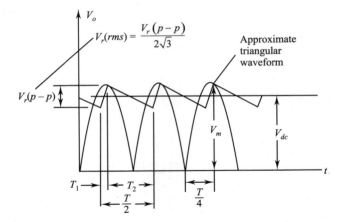

Fig. E1 Approximate triangular ripple voltage for capacitor filter.

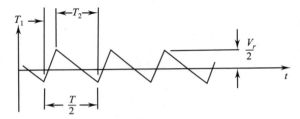

Fig. E2 Ripple voltage.

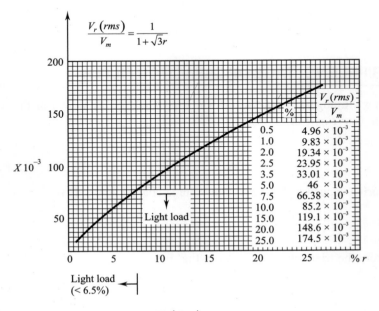

Fig. E3 Plot of $\dfrac{V_r(rms)}{V_m}$ as a function of % r.

Appendix-F

HYBRID PARAMETERS—GRAPHICAL DETERMINATIONS AND CONVERSION EQUATIONS (EXACT AND APPROXIMATE)

F.1 GRAPHICAL DETERMINATION OF THE *h*-PARAMETERS

Using partial derivatives (calculus), it can be shown that the magnitude of the *h*-parameters for the small-signal transistor equivalent circuit in the region of operation for the common-emitter configuration can be found using the following equations.*

$$h_{ie} = \frac{\partial v_i}{\partial i_i} = \frac{\partial v_{be}}{\partial i_b} \cong \left.\frac{\Delta v_{be}}{\Delta i_b}\right|_{V_{CE}=\text{constant}} \qquad \text{(ohms)}$$

$$h_{re} = \frac{\partial v_i}{\partial v_0} = \frac{\partial v_{be}}{\partial v_{ce}} \cong \left.\frac{\Delta v_{be}}{\Delta v_{ce}}\right|_{I_B=\text{constant}} \qquad \text{(unitless)}$$

$$h_{fe} = \frac{\partial i_o}{\partial i_i} = \frac{\partial i_c}{\partial i_b} \cong \left.\frac{\Delta i_c}{\Delta i_b}\right|_{V_{CE}=\text{constant}} \qquad \text{(unitless)}$$

$$h_{oe} = \frac{\partial i_o}{\partial v_o} = \frac{\partial i_c}{\partial v_{ce}} \cong \left.\frac{\Delta i_c}{\Delta v_{ce}}\right|_{I_B=\text{constant}} \qquad \text{(siemens)}$$

In each case, the symbol Δ refers to a small change in that quantity around the quiescent point of operation. In other words, the *h*-parameters are determined in the region of operation for the applied signal so that the equivalent circuit will be the most accurate available. The constant values of V_{CE} and I_B in each case refer to a condition that must be met when the parameters are determined from the characteristics of the transistor. For the common-base and common-collector configurations, the proper equation can be obtained by simply substituting the proper values of v_i, v_o, i_i and i_o.

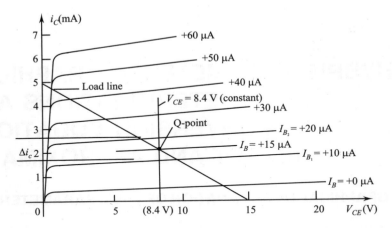

Fig. F.1 h_{fe} determination

The parameters h_{ie} and h_{re} are determined from the input or base characteristics, whereas the parameters h_{fe} and h_{oe} are obtained from the output or collector characteristics. Since h_{fe} is usually the parameter of greatest interest, we shall discuss the operations involved with equations, such as Eqs. (F.1) through (F.4), for this parameters first. The first step in determining any of the four hybird parameters is to find the quiescent point of operations as indicated in Fig. F.1. In Eq. (F3) the condition V_{CE} = constant requires that the changes in base curent and collector current be taken along a vertical straight line drawn through the Q-point representing a fixed collector-to-emitter voltage. Equation (F.3) then requires that a small change in collector current be divided by the corresponding change in base current. For the greatest accuracy, these changes should be made as small as possible.

In Fig. F.1, the change in i_b is chosen to extend from I_{B1} to I_{B2} along the perpendicular straight line at V_{CE}. The corresponding change in i_c is then found by drawing the horizontal lines from the intersections of I_{B1} and I_{B2} with V_{CE} = constant to the vertical axis. All that remains is to substitute the resultant changes of i_b and i_c into Eq. (F.3). That is,

$$|h_{fe}| = \left.\frac{\Delta i_c}{\Delta i_b}\right|_{V_{CE}=\text{constant}} = \left.\frac{(2.7-1.7)\,\text{mA}}{(20-10)\,\mu\text{A}}\right|_{V_{CE}=8.4\,\text{V}}$$

$$= \frac{10^{-3}}{10 \times 10^{-6}} = 100$$

In Fig. F.2, a straight line is drawn tangent to the curve I_B through the Q-point to establish a line I_B = constant as required by Eq. (F.4) for h_{oe}. A change in V_{CE} was then chosen and the corresponding change in i_c determined by drawing the horizontal lines to the vertical axis at the intersections on the I_B = constant line. Substituting into Eq. (F.4), we get

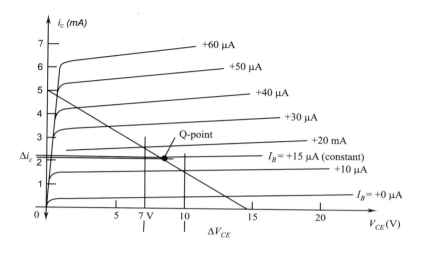

Fig. F.2 h_{oe} determination.

$$|h_{oe}| = \left.\frac{\Delta i_c}{\Delta v_{ce}}\right|_{I_B=\text{constant}} = \left.\frac{(2.2-2.1)\,\text{mA}}{(10-7)\,\text{V}}\right|_{I_B=+15\mu A}$$

$$\frac{0.1\times 10^{-3}}{3} = 33\ \mu A/V = 33 \times 10^{-6} S = 33\ \mu S.$$

To determine the parameters h_{ie} and h_{re} the Q-point must first be found on the input or base characteristics as indicated in Fig F.3. For h_{ie}, a line is drawn tangent to the curve V_{CE} = 8.4 V through the Q-point to establish a line V_{CE} = constant as required by Eq (F.1). A small change in V_{be} is then chosen, resulting in a corresponding change in i_b Substaituting into Eq. (F.1), we get

$$|h_{ie}| = \left.\frac{\Delta v_{be}}{\Delta i_b}\right|_{V_{CE}=\text{constant}} = \left.\frac{(733-718)\,\text{mV}}{(20-10)\,\mu A}\right|_{V_{CE}=8.4\,V}$$

$$= \frac{15\times 10^{-3}}{10\times 10^{-6}} = 1.5\ k\Omega$$

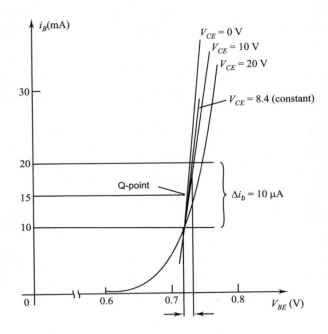

Fig. F.3 h_{fe} determination.

The last parameter, h_{re}, can be found by first drawing a horizontal line through the Q-point at $I_B = 15$ μA. The natural choice then is to pick a change in V_{CE} and find the resulting change in V_{BE} as shown in Fig F.4.

Substituting into Eq.2, we get

$$|h_{re}| = \left.\frac{\Delta v_{be}}{\Delta v_{ce}}\right|_{I_B = \text{constant}} = \frac{(733-725)\,mV}{(20-0)} = \frac{8 \times 10^{-3}}{20} = 4 \times 10^{-4}$$

For the transistor whose characteristics appear in Figs F.1 through F.4, the resulting hybrid small-signal equivalent circuit is shown in Fig F.5.

As mentioned earlier, the hybrid parameters for the common-base and common-collector configuration can be found using the same basic equations with the proper variable and characteristics.

Table F1 lists typical parameter values in each of the three configurations for the broad range of transistors available. The minus sign indicates that in Eq. (F3) as one quantity increases in magnitude within the change chosen, the other decreases in magnitude.

Appendix

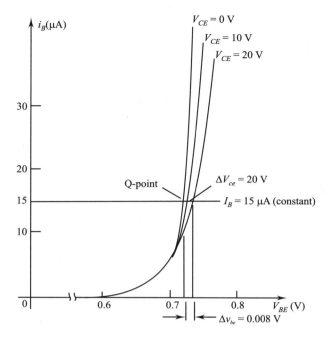

Fig. F.4 h_{re} determination.

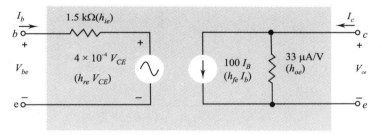

Fig. F.4 Complete hybrid equivalent circuit for a transistor having the characteristics that appear in Figs. F.1 through F.4.

Table F.1 Typical Parameter Values for the CE, CC, and CB Transistor Configurations

Parameter	CE	CC	CB
h_i	$1 k\Omega$	$1 k\Omega$	20Ω
h_r	2.5×10^{-4}	≈ 1	3.0×10^{-4}
h_f	50	-50	-0.98
h_o	$25 \mu A/V$	$25 \mu A/V$	$0.5 \mu A/V$
$1/h_o$	$40 k\Omega$	$40 k\Omega$	$2 M\Omega$

F.2 EXACT CONVERSION EQUATIONS

Common-Emitter Configuration

$$h_{ie} = \frac{h_{ib}}{(1+h_{fb})(1-h_{rb}) + h_{ob}h_{ib}} = h_{ic}$$

$$h_{re} = \frac{h_{ib}h_{ob} - h_{rb}(1+h_{fb})}{(1+h_{fb})(1-h_{ib}) + h_{ob}h_{ib}} = 1 - h_{re}$$

$$h_{fe} = \frac{h_{fb}(1-h_{rb}) - h_{0b}h_{ib}}{(1+h_{fb})(1-h_{rb}) + h_{ob}h_{ib}} = -(1+h_{fc})$$

$$h_{oe} = \frac{h_{ob}}{(1+h_{fb})(1-h_{rb}) + h_{ob}h_{ib}} = h_{oc}$$

Common-Base Configuration

$$h_i = \frac{h_{ie}}{(1+h_{fe})(1-h_{re}) + h_{ie}h_{oe}} = \frac{h_{ic}}{h_{ic}h_{oc} - h_{fc}h_{rc}}$$

$$h_{rb} = \frac{h_{ie}h_{oe} - h_{re}(1+h_{fe})}{(1+h_{fe})(1-h_{re}) + h_{ie}h_{oe}} = \frac{h_{fc}(1-h_{rc}) + h_{ic}h_{oc}}{h_{ic}h_{oc} - h_{fc}h_{rc}}$$

$$h_{fb} = \frac{-h_{fe}(1-h_{re}) - h_{ie}h_{oe}}{(1+h_{fe})(1-h_{re}) + h_{ie}h_{oe}} = \frac{h_{rc}(1+h_{fc}) - h_{ic}h_{oc}}{h_{ic}h_{oc} - h_{fc}h_{rc}}$$

$$h_{ob} = \frac{h_{oe}}{(1+h_{fe})(1-h_{re}) + h_{ie}h_{oe}} = \frac{h_{oc}}{h_{ic}h_{oc} - h_{fc}h_{rc}}$$

Common-Collector Configuration

$$h_{ic} = \frac{h_{ib}}{(1+h_{fb})(1-h_{rb}) + h_{ob}h_{ib}} = h_{ie}$$

$$h_{rc} = \frac{1+h_{fb}}{(1+h_{fb})(1-h_{rb}) + h_{ob}h_{ib}} = 1 - h_{re}$$

$$h_{fc} = \frac{h_{rb}-1}{(1+h_{fb})(1-h_{rb}) + h_{ob}h_{ib}} = -(1+h_{fe})$$

$$h_{ob} = \frac{h_{ob}}{(1+h_{fb})(1-h_{rb}) + h_{ob}h_{ib}} = h_{oe}$$

F.3 APPROXIMATE CONVERSION EQUATIONS

Common-Emitter Configuration

$$h_{ie} \cong \frac{h_{ib}}{1+h_{fb}} \cong \beta r_e$$

$$h_{re} \cong \frac{h_{ib}h_{ob}}{1+h_{fb}} - h_{rb}$$

$$h_{fe} \cong \frac{h_{fb}}{1+h_{fb}} \cong \beta$$

$$h_{oe} \cong \frac{h_{ob}}{1+h_{fb}}$$

Common-Base Configuration

$$h_{ie} \cong \frac{h_{ie}}{1+h_{fe}} \cong \frac{-h_{ic}}{h_{fc}} \cong r_e$$

$$h_{rb} \cong \frac{h_{ie}h_{oe}}{1+h_{fe}} - h_{re} \cong h_{re} \cong 1 - \frac{h_{ic}h_{oc}}{h_{fc}}$$

$$h_{fb} \cong \frac{h_{fe}}{1+h_{fe}} \cong -\frac{(1+h_{fc})}{h_{fc}} \cong -\alpha$$

$$h_{ob} \cong \frac{h_{oe}}{1+h_{fe}} \cong \frac{-h_{oc}}{h_{fc}}$$

Common-Collector Configuration

$$h_{ic} \cong \frac{h_{ib}}{1+h_{fb}} \cong \beta r_e$$

$$h_{rc} \cong 1$$

$$h_{fc} \cong \frac{-1}{1+h_{fb}} \cong -\beta$$

$$h_{oc} \cong \frac{h_{ob}}{1+h_{fb}}$$

Appendix-G

SELECTED TRANSISTOR

Characteristics ....

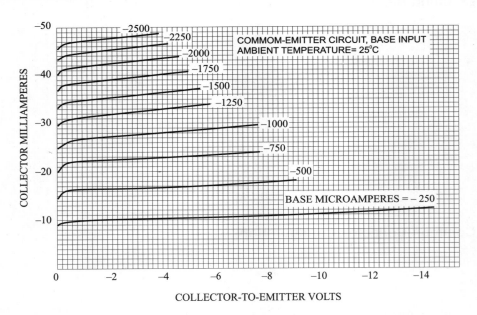

Fig. G1 Average collector characteristics of RCA transistor 2N104 in common-emitter mode.

Appendix

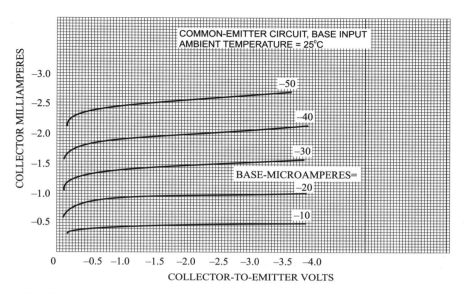

Fig. G2 Average collector characteristics of RCA transistor 2N104 in common-emitter mode, high current.

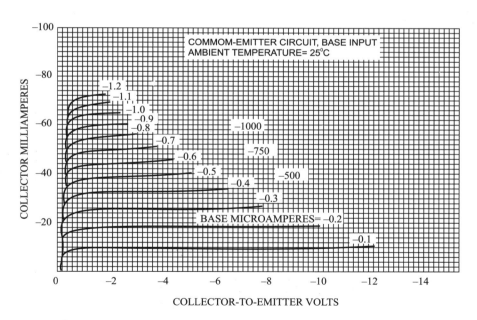

Fig. G3 Average collector characteristics, common-emitter connection, RCA transistor 2N139.

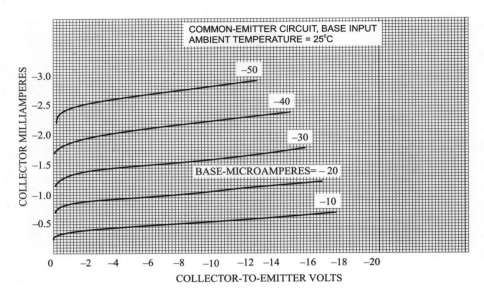

Fig. G4 Average collector characteristics, common-emitter connection, RCA transistor 2N139.

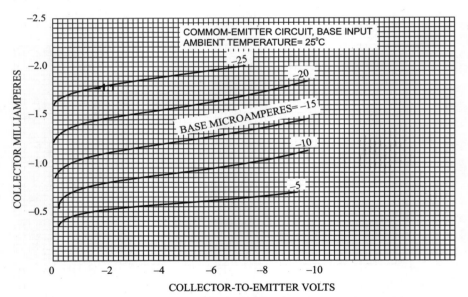

Fig. G5 Average collector characteristics, common-emitter mode, RCA transistor 2N175.

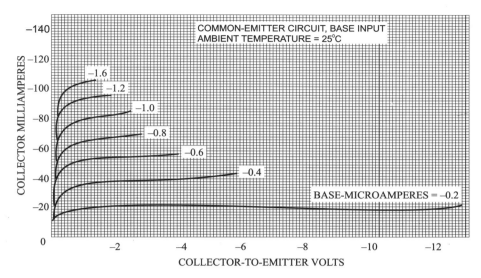

Fig. G6 Average collector characteristics, common-emitter mode, RCA transistor 2N270.

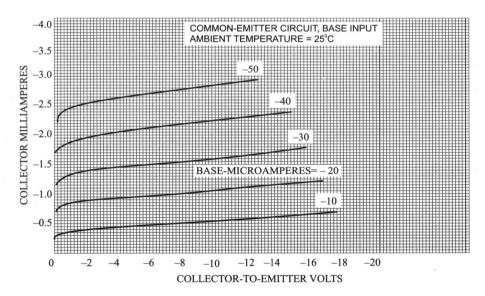

Fig. G7 Average collector characteristics of RCA transistor 2N410 in common-emitter mode.

GLOSSARY OF ELECTRIC TERMS

AC — Alternating current. In a time/voltage diagram, AC voltage represents a sine function (usually), or just any periodically alternating function. The mains voltage is AC voltage, for example.

Active high/low — Normally, signals are active high, which means a voltage level of 0V represents a logical 0 (LOW) and a voltage of above 5V represents a logical 1 (HIGH). If, for example, an IC pin is named "CS" (chip select), the chip is usually selected by pulling this line to HIGH (5V for TTL), and it gets deselected by pulling it to LOW (0V).

ADC — Analog-Digital Converter.

Ammeter — Device for measuring electric current. Usually part of a multi meter.

AND — Logical function which is TRUE if all inputs are TRUE.

A B	A AND B
0 0	0
0 0	0
1 0	0
1 1	1

Examples:

7408 : 4 AND gates with 2 inputs each

7409 : 4 AND with 2 inputs each, open collector

4081 : 4 CMOS and gates with 2 inputs each.

BGA	Ball Grid Array. A type of chip package where the fixing method consists of a number of solder balls mounted under the chip and directly soldered onto a PCB.
Bread board	Board made of pertinax or other insulating material for building prototype circuits. It contains a matrix of holes. There are also types with soldering pads around the holes, these cost more but are easier to work with.
Buffer	Same as driver.
Bus	The name of a set of lines/signals fulfilling a common function, *e.g.* the address bus and the data bus. Examples include the PCI bus, H.100 and H.110 buses.
BASE	The input terminal of a bipolar transistor.
BETA	The Greek letter that designates the current gain of a bipolar transistor. It is the ratio of the transistor's output current (IC) to its input current (IB).
BIAS Voltage	The DC voltage applied across the terminals of a PN junction, whether the device is a diode, bipolar transistor, or JFET. A PN junction is forward biased when a positive voltage is applied to the P-region with respect to the N-region, and reversed biased when the voltage polarity is reversed.
Bipolar Transistor	A three-terminal semiconductor component with a three-layer structure of alternate negative and positive type materials (NPN or PNP). It provides current gain and voltage amplification in a circuit.
Bridge Rectifier	Four semiconductor diodes configured as a bridge that acts to change AC to full-wave pulsating DC.
Cathode	One of the two terminals of a diode (negative type material) or the terminal (also negative type material) that is common to both input and output sections of an SCR.
Chips	Unpackaged diodes, bipolar transistors, SCRs, TRIACs, and field-effect transistors (FETs) - also called DICE.
CMOS (Complementary Mosfet)	A combination of an N-channel and a P-channel MOSFET in a single switching circuit. This circuit features very low power dissipation and the effective elimination of an external load resistor. The device responds to a digital pulse at its input by turning one section of the device ON and the other OFF, causing the turned OFF section to act as its high-resistance load. When the input pulse reverts to zero, the state of the two sections of the device are reversed.

Glossary

Collector — The output terminal of a bipolar transistor.

Complementary Bipolar transistors — An arrangement of NPN and PNP bipolar transistors in which the polarity of the supply voltage applied to one device is the reverse of the other. The two transistors normally have identical electrical characteristics and are used as a matched pair.

Capacitance — Electrical entity which describes the amount of charge a capacitor can store unit farad (F). A capacitor is an electrical element which is capable of storing small amount of electrical energy, just like an accumulator. The five most common capacitor types are:

Styroflex capacitor — High quality, little tolerance. Mainly employed in high-end audio applications. Irrelevant for computer applications. Un-polarized.

electrolytic capacitor — High capacities, polarized, bigger tolerances. Typical application: filtering capacitor in power supplies. Typical capacity greater than 1µF.

Ceramic capacitor — Un-polarized. Typical capacity smaller than 1µF. The Dielectric consists of ceramic layers. Widely used for all applications.

Film capacitor — Like ceramic, self-healing, usually smaller tolerance range, as ceramic, this type is widely used in all applications. Un-polarized. Available for high voltages also (up to 1000V).

Tantalum (electrolytic) capacitor — Like electrolytic, smaller tolerance range, particularly used in digital electronics. Polarized. Typical application: stabilizing. Rarely available for higher voltages (>10V) and higher capacities (>100 µf) or at least very expensive then.

Since a simple capacitor only consists of two plates facing other, you can imagine that even two wires lying in parallel have a certain capacitance. When you charge a capacitor by applying voltage to it, it first behaves like a shortcut, then its resistance increases until no current flows through it anymore. This shortcut period is also present in parallel wires (*e.g.* a cable), it drains lots of power from the chip the wires are connected to and the longer the cable, the higher its capacitance, the longer the shortcut period, the higher the current which the chip has to endure, and the shorter the chip's lifetime.

Chip	Generic Term. An IC in a housing or package. Package types may be Thru-Hole (THT), Surface Mount (SMT/SMD), Ball Grid Array (BGA) or Wafer level Chip-scale Packaging (WLCSP).

Thru-hole   SMT/SMD   BGA

Dimensions and sizes for chips are defined by JEDEC.

The following types are some examples:

DIL/DIP	Dual In Line (DIL). This is the most widely used IC housing. The pins come out on both sides of the chip. When the notch on the case points to the top, pin 1 is in the upper left corner, the other pin numbers are counted counter-clockwise. Also used in 'DIP switch', a set of small switches in a chip like case.
SIL/SIP	Single In Line (SIL). They have pins on only one side of the case. SILs are used on SIMMs (Single In Line Memory Module) and SIPs (single In line Peripheral package).
SOP, SOT, SOIC, TSOP	Examples of surface mount package (SMT) types. May be Dual-in-line or have pins on all four sides.

etc.

BGA, FBGA etc Examples of Ball Grid Array types etc.

CMOS	Complementary Mental Oxide Semiconductor. TTL uses bipolar transistors, while CMOS chips use uni-polar transistors (FETs) which are connected complementarily (one p-mos, one n-mos) thus consuming virtually no power and staying much cooler than appropriate TTL chips. Alas, CMOS chips are not suitable for very high frequencies: when the input level changes, the supply voltage pin and GROUND get quickly short-circuited. The higher the switch frequencies, the higher the shortcut time. If you run CMOS chips at high frequency, most of the switching time there is a shortcut, resulting in high power consumption and heat generation.

The switching thresholds are less than 30% (LOW) and greater than 70% (HIGH) of the supply voltage.

As opposed to the TTL series (74xx), CMOS family chips are not bound to 5V supply voltage. V_{cc} ranges from 3 to 18V (for the 4000 family). There are also TTL compatible CMOS families available, e.g. the widely used 74 HCxx series (voltage range 4-6V), where HC stands for high speed Cmos.

Glossary

Composite video — Video signal which comprises of color and brightness information as well as horizontal and vertical synchronization information. Since the video chip's output signals are mixed into one signal (the composite video signal) and then must be split again in the monitor, losses occur and deteriorate the display quality, often resulting in color streaks. If possible, use the computer's Chroma/Luma output, which carries brightness and sync information on one line, but color information on another line, which eliminates the color streaking. The best result is achieved by using an RGB output.

Conductor — A materials is called a conductor if electrons can move through it, in other words, if it allows flow of electrical current. How well current can flow through the conductor is determined by its resistance. If the resistance is very high, the material is called an insulator.

Connector — Many types of connectors are used the following list indicates some of the most common:

BNC — Bayonet Nut Connector (you may also see it spelled as Bayonett if you are German, Bayonette if you are French or Bayonett if you are Spanish - so now you know). Used for video connections, Ethernet (10 base2)/arcnet, and for high frequencies *e.g.* measuring equipment (oscilloscope, etc. and RF applications).

DB-xx — Used for : RS232 C (DB9 or DB25 male), parallel port (DB25 female). For some pinouts.

DIN — Deutsches Institute fur Normung. Used for AT style PC keyboards (5-Pin), PS/2 mice (6-Pin mini-DIN) ATX style keyboards (6-pin mini-DIN) and for MIDI connections. For pinouts here.

RCA [Cinch] — Radio Company of America. Used for audio and video connections. In Germany and probably other countries, too, this connector is also known as 'cinch'.

SMA/SMB — RF Co-ax connector.

TNC — RF Co-ax connector. May have standard or reverse polarity (mandated by FCC for use with ISM band radios *e.g.* 802.11).

TNC connector polarity

	Jack	Plug
Standard	female	male
Reverse - Polarity	male	female

N-Type — RF Co-ax connector.

Continuity — A cable (or other conducting material) has continuity when it has a low resistance, when it therefore constitutes a shortcut.

Continuity tester — Device for checking for continuity. It reacts to a resistance below ~100 ohm, normally acoustically; some devices have a selectable threshold. Usually part of a multimeter.

Counter — Counters are elements counting the number of clock signals and outputting them as binary or decimal representation on the output pins.

Examples:

4060: 14-step CMOS binary counter with internal oscillator circuit 7468: 2 asynchronous decimal counters.

Current — Electrical entity which is defined by he amount of charge flow in Coulomb per second. Unit: = Ampere (A). 1 A = 1C / 1 Sec. Symbol: = I in electrical equations.

DC — Direct current. DC voltage is linear and constant, and either positive or negative. The same applies to direct current. See also AC.

DAC — Digital-Analog Converter.

Diode — Semiconductor element which lets current flow in only one direction (forward direction). Current flows if a positive voltage greater than the forward voltage is applied to the anode of the diode (the other end is called cathode and is usually marked with a black ring on the case), otherwise, the diode has a very high resistance. If the applied voltage is below the avalanche/blocking voltage (which is always negative), the diode breaks down and constitutes a shortcut.

The rarely used germanium diodes have a forward voltage of 0.3V, while the standard silicon diode has a forward voltage of 0.6V. (Graphic symbol)

> *Schottky diode* — Diode with a P-N junction consisting of metal and silicon [?]. It is used for applications requiring fast switching, for instance ECL circuits.
>
> *Zener diode [z-Diode]* — As opposed to all other diodes, the Z diode is used in reverse direction. It has a defined avalanche voltage and is often used for voltage stabilizing.
>
> Z diodes often have a blue, yellow or red base colour. Common series are BZXxx, ZPDxx, and BZYxx, where xx is the avalanche voltage, *e.g.* ZPD4.7 or BZY9.1.

Glossary

	Tunnel diode Only for very high frequency applications. Its function is not based on the avalanche effect, but on the tunnel effect.
DRAM	Dynamic RAM. DRAM needs a continuous refresh (through the use of CAS and RAS signals), as the information in it is stored by very small capacitors.
Driver	Sometimes called a Buffer. A driver's output level follows the input level if it is a non-inverting type, and it implements a NOT function, if it is an inverting type. Drivers are employed for

- increasing the maximum output current of logical signals.
- signal shaping - turn a noisy signal into a clean signal
- protecting expensive chips

Examples:

7404: hex inverter (6 inverters)

7414: hex inverter with schmitt trigger inputs.

7405: hex inverter with O.C. outputs

7406: Inverting driver with O.C. outputs (30V)

7416: inverting driver with O.C. outputs (15V)

7407: non-inverting driver with O.C. outputs (30V)

4069: inverting CMOS driver

4049: inverting CMOS driver, buffered

4050: CMOS driver

ECL	Emitter-Coupled-Logic. Very fast logic family and used in some processor designs such as the AMD2900 range.
EEPROM	Electrically Erasable PROM. In contrast to EPROMs, EEPROMs don't need exposure to UV light to be erased, but can be erased electrically. A big advantage is that is accessed like an SRAM. Write accesses perform an automatic clear before write and thus make writing EEPROMs as easy as writing to SRAMs. Series designator: 28xx, where xx is the number of where bits stored.
	Serial EEPROMs are labeled 24Cxx (8 bit) or 93Cxx (16 bit).
	Low voltage eproms (PLCC): 3.3 V are labeled 27Vxx.
Electron	To be supplied.
EPROM	Erasable PROM. EPROMs allow the contents to be erased by exposing its built in window to UV light. After this process, all memory cells contain $ff and the EPROM can be written again.
	Series designator: 27xx, where xx is the number of K bits stored.
EXOR (XOR)	Exclusive OR. Logical function which is TRUE, if and only if, exactly one input is TRUE. Frequently called XOR.

A B	A XOR B
0 0	0
0 1	1
1 0	1
1 1	0

Examples:

7486: 4 XOR gates with 2 inputs each

74136: 4 XOR gates with 2 inputs each, open collector

4070: 4 CMOS XOR gates with 2 inputs each

FET — Field Effect Transistor. As opposed to normal bipolar transistors, these unipolar transistors have a negligible flow of current through their gate (bipolar: base), they consume virtually no power. NMOS-FETs and PMOS-FETs can be coupled to form CMOS circuits.

FLASH — An EEPROM which can be written (and erased) in whole banks or sectors. Typically comes un Uniform sectored or Bootstrap Sectored designs. Series designator: 28Fxxx (12 V prg. voltage), 29Fxxx (5 V prg. voltage), and 29LVxxx/29SLxxx (3 V and below), where xxx is the memory capacity: 010-1 Mbit, 020 - 2 Mbit, etc. If the Flash supports 16 bit organization, xxx is: 100 - 1 Mbit, 200 - 2 Mbit, etc. Early Flash (before 1998) had a limit on the number of erase cycles (typically 100,000) but most modern FLASH has essentially no erase limits.

Flip-flop — This edge-triggered element has two stable states, which are toggled on different events, depending on the type:

D flip-flop — Delay flip-flop. The input is copied to the output delayed by one clock cycle. D-type flip-flops are normally positive (rising) edge triggered but both edge types are available.

T flip-flop — Toggle flip-flop. The output alternates with each input signal change. To simulate a T flip-flop, you can simply connect a D flip-flop's complementary output Q with its input.

JK flip-flop — This type combines characteristics of RS flip-flop and T flip-flop. It has two inputs J and K and a clock input C. If different signals are applied to J and K, the JK flip-flop acts like an RS flip-flop. If J = K, it acts like a T flip-flop.

Viewed technically, a JK flip-flop comprises of two coupled flip-flops (called Master and Slave), where one outputs the input signals on the rising edge, the other one on the falling edge of the clock signal. Therefore, it is sometimes called master slave flip-flop. Main applications of the JK flip-flop are counters and shift registers. J-K flip-flops are normally negative (falling) edge triggered but both edge types are available.

RS flip-flop   Reset-Set flip-flop. These have a reset input and a set input and a set input. If reset is high, the output goes low, if set is high, the output goes high. Setting both reset and set to high is forbidden, as the results are indetermined.

Examples:

7470: JK flip-flop with 3 inputs each, preset and reset.

74L71: RS master slave flip-flop with 3 inputs each, preset and reset.

74171: 4 D flip-flops with clear input.

Float   An Electronic signal is said to 'float' when its value is not defined under all conditions. Floating is generally a 'bad thing' since random effects (*e.g.* induction) could easily change the value with unexpected or unpleasant results. Signals that would otherwise 'float' are typically 'pulled-up' (high) or 'pulled-down' (Low) with a weak resistor such that they can be easily changed when driven.

Fuse   A device designed to break a circuit when too much voltage or current is applied. The idea being that its cheaper to replace a fuse than a device.

Electronic fuses:

Domestic fuses: Historically little glass tubes with a wire of defined maximum voltage and current which melts when its capacity is exceeded. There are two common formats: 5 × 20 mm (German) and 6 × 30 mm (American). Modern wiring typically uses circuit-breakers which can be reset rather than replaced.

Gate   A gate is a circuit on a chip, which implements a logical function. A 7406, for example, contains 6 gates (non-inverting drivers).

IC   Generic Term. Integrated Circuit. A set of gates etched on a silicon wafer. As ICs are very sensitive, they are enclosed or packaged in a plastic or ceramic case/carrier, with their inputs and outputs connected to metal pins or balls. An IC in a package is commonly referred to as a CHIP. Chips are also called ICs!

Impedance	Expressed in ohms is vector sum of all opposition to the flow of current in a (typically AC) circuit which includes resistance, capacitance and inductance.
Inductance	Measures in Henries. The ability of a component to store energy in the form of a magnetic field.
Inductor	A passive device that stores electrical energy in the form of a magnetic field. Normally consists of a wire loop or coil. Inductors are typically used to smooth out voltage fluctuations in power supply circuits.
Insulator	A martial which doesn't conduct electrical current. The opposite in a conductor.
Inverter	Gate inverting a logical signal, thus implementing a NOT function. For examples, see drivers.
Latch	A set of flip-flops with a common clock signal. In each cycle, they take the logical input signals over to their outputs. Usually used to form multiplex address busses. As opposed to flip-flops, latches are level-triggered.
LED	An LED (Light Emitting Diode) is a diode emitting light when operated in a forward direction. Since it is a diode, it has a nearly negligible resistance and MUST be operated with a series resistor. The forward voltages depend on the type: Red 1.6V - 2.1V Series resistor for 5V: 330 ohm. Green 2.2V - 2.7V. Series resistor for 5V: 270 ohm Yellow 2.7V - 3.2V. Series resistor for 5V: 140 ohm White 3.3V - 4.2V. Series resitor for 5V: 75 ohm Blue 3.3V - 4.2V. Series resistor for 5V: 75 ohm. While normal LEDs consume about 20 mA, high efficiency LEDs require only currents from 2-4mA (depending on type and color), which means that you can directly connect them to standard logical output (74LS xx or CMOS 4000 series) without the need for a driver. Nevertheless you still need an appropriate series resistor. Resistor calculation = voltage drop × current required in amps.
Logic Tester/ Probe	Detects and indicates logic TTL (and/or) CMOs voltage levels. It usually contains a pulse memory (comprising a flip-flop) that memorizes pulses too short to be noticed otherwise.
Mains voltage	The voltage at the wall outlet. Australia: 240V @ 50Hz UK: 230V @ 50 Hz

Germany: 230V / 400 V @ 50 Hz (formerly 220 V/ 380V)
Japan: 100 V @ 75 Hz
USA: 120V/125V @ 60 Hz

Note that since 1989, the standard European voltage is 230 V @ 50Hz.s

Monoflop   Also known as one-shot multivibrator. Flip-flop with only one stable state. It remains in the unstable state for a certain time determined by capacitors.

Examples:

74121: Monoflop with schmitt trigger input 74221: 2 monoflops with schmitt trigger input and reset 74122: Retriggerable monoflop with reset 74123: 2 retriggerable monoflops with reset.

MOS   Metal Oxide Semiconductor.

Multi-meter   An all-in-one measuring device. It combines a volt-meter, an amp-meter and an ohm-meter which usually also can act as continuity tester. Often it contains a transistor tester and measures capacities and inductivities (in a small range). There are both analog and digital types, the latter is the preferred choice.

NAND   Logical function which is TRUE if and only if not all of the inputs are TRUE.

A	B	A NAND B
0	0	1
0	1	1
1	0	1
1	1	0

Examples:

7400: 4 NAND gates with 2 inputs each

7401: 4 NAND gates with 2 inputs each, open collector

4012: 2 CMOS NAND gaes with 4 inputs each

4093: 4 CMOS NAND gates with 2 inputs each and schmitt trigger.

Negative logic   Negative logic means that the signals are active low.

NMOS   N-doped MOS

NOR   Logical function which is TRUE if and only if all inputs are FALSE.

A	B	A NOR B
0	0	1
0	1	0
1	0	0
1	1	0

Examples:

7402: 4 NOR gates with 2 inputs each

7423: 4 NOR gates with 4 inputs each, open strobe

4001: 4 CMOS NOR gates with 2 inputs each

4002: 2 CMOS NOR gates with 4 inputs each

NOT   Logical function which is TRUE if the input is FALSE.

A	NOT A
0	1
1	0

Ohm's Law   Defines the relationship between voltage (E) current (I) and Resistance (R) in a circuit. For DC circuits Ohms law is:

I = E / R (amps = volts / resistance in ohms)

Or

E = I × R (volts = amps × resistance in ohms)

Additional equations.

Ohmmeter   Device for measuring resistance. Usually part of a multi-meter.

Open collector   A possible output connection of a TTL circuit. The output is formed by a single transistor, which is not connected to the supply voltage, therefore an external connection to the supply voltage (via a pull-up resistor) is required. Multiple open collector outputs can be connected together, the output carrying a 0 signal will override all other outputs.

Oscilloscope   A test device which displays voltage curves graphically.

OR   Logical function which is TRUE if at least one input is TRUE.

A	B	A OR B
0	0	0
0	1	1
1	0	1
1	1	1

Examples:

7432: 4 OR gates with 2 inputs each

74832: 6 OR drives with 2 inputs each

4071: 4 CMOS OR gaes with 2 inputs each

4072: 2 CMOS OR gates with 4 inputs each

PAL	This acronym has two meanings: 1. Phase-Alternation Lines. Video encoding standard used in European countries. PAL has 50 pictures/sec interlaced and a resolution of 625 lines [?]. 2. Programmable Array Logic. A chip which implements a sum-of-products logic equation. A PAL can be programmed only once. Type designator: xxyzz, where xx is the number of inputs, y is either L for active low outputs or H for active high outputs, and zz is the number of outputs; example: 16L8. A derivate [?] is the PLA.
PCB	Printed Circuit Board. The circuit tracks or traces are etched photographically onto a media. PCBs may be single-sided (tracks on one side only), double-sided (both top and bottom surfaces are used) or multi-layer where tracks are placed on a number or separate layer which are then bonded together. Tracks are connected on multi-layer boards using VIAs (small holes). Holes are drilled in the board for thru-hole technology or solder pads provided for SMT or BGA devices. Components may be placed on the top or increasing on both the top and bottom of a PCB.
Photo diode	Diode which is controlled by light.
Photo transistor	Transistor which is controlled by light.
PLA	Programmable Logic Array. The same as a PAL, but with a programmable OR matrix.
PMOS	P-doped MOS
P-n Junction	
Positive logic	Positive logic means that the signals are active HIGH. Negative logic means that signals are active LOW (Most commonly in RS 232 circuits)
Potentiometer	A variable resistor the value of which is determined by the position of a slider or a knob.
PROM	Programmable ROM. This memory type can be written once, then it behaves like a ROM. Series designator: 25xx, where xx is the number of kbits stored.
Pull-up/pull down resistor	Pull-ups (or pull-downs) have two primary purposes both of which are variations on a fundamental theme which is to prevent a short-circuit by adding a resistor in the path between V_{cc} and GND for a particular signal. **Configuration:** Many ICs have pins which must be set to a HIGH or LOW to configure the chip. Unless are IC is defined to have an internal pull-up or pull-down you typically use a pull-up (the resistor is between the signal pin and V_{cc}) to set a

HIGH (1) or a pull-down (the resistor is between the signal pin and GND) to set a LOW (0).

Floating Signals: If a signal is not being actively driven all the time it will float (*i.e.* take an arbitrary and maybe changing value). To prevent this it may be pulled-up (HIGH) or pulled-down (LOW) into a default state. Pull-ups or pull-downs are usually weak (*i.e.* high value resistors of 4.7K, 10K (most common) or 47K) since in the case of floating signals this allows the 'driven' level to overcome the resistance with a modest current. For minimum power loss especially in configurtion function use the highest value (47K). Since higher resistance values take longer to overcome than lower values if the signal needs to be stable very quickly you may need to go as low as 1 K for the pull-up (pull-down.)

RAM	Random Access Memory. Information can be read and written in any order, the number of read or write accesses is not limited. RAM comes in different flavors: DRAM, SRAM, SDRAM, EDO-RAM, VRAM and many more.
Rectifier	Circuitry transforming AC into DC, usually consisting of 4 diodes (aka bridge rectifier)
Resistance	The resistance of a conductor (or an insulator) is how easily current can flow through it. Unit: ohm (capital omega) Symbol = R.
Resistor	Electrical element with a defined resistance. It is used as voltage divider, current limiter or for ensuring that signals do not float. For small through-hole resistors, their value is not printed on the case, but encoded with color rings.
Radio Frequency	Generic term defines equipment which works in the radio frequency range typically ? to ?
RGB	Red-Green-Blue. These three colors are additively mixed in color TVs and monitors and so give a picture which ranges from black over all rainbow colors to white. The number of colors displayed depends on the technology: TTL or ECL supply digital signals and thus a limited color resolution, usually 4 bits, which results in 16 colors; analog signals, however, make the color resolution practically infinite, the number of colors only depends on the graphics card's memory and on its RAMDAC or VRAM.
RMS	Root Mean Square. The real peak value of an AC voltage, which is U* square root of 2, abbreviation V_{rms}.
ROM	Read only memory. Unlike RAM, this type of electronic memory can only be read. The ROM's content is determined during the manufacturing process (mask programming). Derivatives are PROM, EPROM, EEPROM, and Flash-EPROM.

SDRAM	Synchonous DRAM. Differs from conventional DRAM in that it internally gates (synchronises) all access using a single clock rather than separate column and row clock (driven by CAS & RAS).
Schmitt trigger	A logical device that outputs 0 if the input voltage is below a given threshld voltage and 1 otherwise. Used to clean up the edges of digital signals. Often comes with a built-in inverter.
Semiconductor	Pure semiconductor materials like silicon are insulators. But doping these materials with a very small amount of *e.g.* Bor makes them less insulating and, under certain circumstances, conduct electrical current. Common semiconductors are diodes and transistors, which are also etched into the silicon wafers of ICs.
SMD or SMT	Surface mounted device (Surface mounted technology). A chip packaging technique. SMD technique means soldering elements (which have specially designed, very short pins) directly onto pads on the PCB surface without drilling holes. Other packaging techniques are 'Thu-hole' and Ball Grid Array (BGA).
Solder	Solder is made of tin (Sn) and lead (Pb) and contains a rosin core, which makes the solder flow more easily.
Soldering iron	A tool for soldering electrical conducting connections.
SRAM	Static RAM. As opposed to DRAM, this type of memory does not need a continuous refresh, as the information in it stored by flip-flops.
Three-state	See tristate.
Thru-hole (THT)	A chip packaging technology and requires holes in the PCB through which component pins were inserted and soldered on the reverse side. Through-hole is still widely use for connectors and other components that also have a physical use since the through hole provides a mechanical achoring function (*e.g.* DB25, RJ45 etc). Surface mount versions of these components exists but almost always use one or more mechanical locating holes or pins. Aternate packaging technologies are surface Mount (SMT/SMD) and Ball Grid Array (BGA).
Thyristor	Sometimes called a semiconductor controlled rectifier. It has 3 pins (anode, cathode and gate). When powered and Gate is ON (high) forward current only will flow from the anode to cathode (irrespective of state of the Gate) until it drops below a certain level (called the Holding Current). It can be used to rectify current.
Totem pole	A possible output connection of a circuit. A totem pole consists of two transistors, which are driven complementary. Depending

	on the desired output, only one of the two transistors is conducting. If two totem pole outputs are connected, a shortcut occurs if they different digital signals (0/1 or 1/0).
Transformer	A transformer changes one AC voltage into another AC voltage. It consists of two coils (actually not separate coils, but windings) with a different number of turns, where one coil (transformer primary winding) encloses the other (transformer secondary winding). The current flowing through the transformer primary (the one where the input voltage is applied) invokes a magnetic field which in turn induces a voltage in the transformer secondary, the amount of which is determined by the ratio of the number of turns of the windings.
	A transformer can have more than one secondary, resulting in more than one output voltage.
Transistor	"Transfer Resistor". Invented in 1948 by John Bardeen and Walter Houser Brattain. In principle, this element is an electrically controllable semiconductor resistor. It has three terminals C(Collector), E(Emitter), and B(Base). Basically, when there is no voltage applied to the base, the transistor acts as an insulator and blocks current flow between C and E.
	It is used both as an amplifier and an electronic switch.
Triac	Provides similar functionality to a thyristor but supports bi-directional current flow.
Tristate or three-state	The output lines of tristate circuits can have three states: HIGH, LOW and HIGH IMPEDANCE (HI-Z), where the latter is equivalent to not being connected.
Voltage	Electrical entity which is the cause for current flow. When talking about AC voltages, peak-to-peak voltage means as the name suggests - the absolute amount of voltage between the upper and the lower bound; abbreviation: V_{cc}. Unit Volt (V)
Voltmeter	Device for measuring electrical voltage. Usually part of a multimeter.
VRAM	Video RAM, VRAM is dual-ported, so that you can read and write simultaneously, resulting in a much smaller access time. As the name.

❑❑❑

INDEX

A

Acceptor 13
Active components 2
AND 228
Approximate analysis 84

B

Bias Current 197
Binary Addition 227
Binary Coded Decimal (BCD)
    Numbers 227
Binary Subtraction 227
Binary System 222
Binary to Decimal 224
Binary to Octal 225
Block diagram of a CRO 273
Block diagram of
    a Digital Multimeter 272
Block diagram of a ramp-type DVM 271
Boolean algebra 227
Boolean Algebra Theorems Table 229

C

Canonical Forms 231
Capacitors 4
Cathode Ray Oscilloscope (CRO) 273
Circuit Configurations 72
Collector to Base Bias 86
Colour Coding of Resistors 3
Common Base (CB) Configuration 72
Common Collector (CC) Configuration 73
Common Drain JFET Configuration 163
Common Emitter (CE) Configuration 72
common gate (CG), common source (CS)
    and common dr 156

Common Gate JFET Configuration 156
Common Source JFET Configuration 160
Comparison of Biasing Circuits 86
Concept of Ideal Op-Amp 189
Concept of Pinch-Off and Maximum
    Drain Saturation 151
Conductance 3
Conversion of bases 224
Current gain A_i 92

D

DC bias with voltage feedback or
    collector bias 85
Decimal System 221
Decimal to Binary 224
Decimal to Hexadecimal 226
Decimal to Octal 225
Depletion Layer 15
Depletion Mode Operation 167
Dielectric 4
Differential amplifier circuit 193
Differentiator 195
Digital voltmeters (DVMs) 269
Diode Equivalent Circuit 28
Diode Ratings 29
Diode Resistance 25
Donor 13

E

Electronic charge and current 2
Electronic Circuit Components 2
Electronics instruments 269
Emitter Bias 81
Emitter Bias Circuit 87
Emitter follower circuit 74
Energy Levels 11

Enhancement Mode Operation 167
Exact analysis 83
Extrinsic Semiconductors 13

F

Fixed Bias Circuit 86
Fixed-Bias 80
Fixed-biasing of JFET 154
Forward biased 21
Forward Biasing 15
Forward Resistance 25
Fractional Binary to Decimal 225
Fractional Decimal Number to Binary 224

G

Graphical Analysis of CE Amplifier 87

H

Hexadecimal System 223
Hexadecimal to Octal 226
Hole 13
Hybrid Equivalent Circuit for Common Base (CB) 95

I

Ideal and Practical *V-I* characteristics 24
IGFET 165
Inductors 4
Input Offset Current 197
Input Offset Voltage 197
Input resistance 92
Integrator 196
Intrinsic Semiconductors 12
Inverting Amplifier 189

J

JFET Biasing 154
JFET Connections 156
Junction Field Effect Transistor 148
Junction Field Effect Transistor (JFET) 148

K

K-map 232
Kinetic energy 6

L

Logic Gates and Universal Gates 229

M

Measurement of Current 275
Measurement of Voltage 275
Measurement of Frequency 276
Measurement of Phase Difference 275
Measurements using CRO 275
Metal Oxide Semiconductor Field Effect Transistor 164

N

N-type semiconductor 14
Non-inverting Amplifier 190
NOT 228
Number System 221

O

Octal System 223
Octal to Binary 226
Octal to Decimal 225
Op-Amp 187
Op-Amp Integrated Circuit 187
Op-amp Parameters 197
Open loop gain 189
OR 228
Output resistance 94
Overall Voltage Gain 97

P

P-N Junction 15
P-N Junction and Depletion Layer 15
P-type of semiconductor 14
Parameter Model 90
Passive components 2
Potential barrier 15
Potential-divider Method of Biasing JFET 156
Product of Sums 231

R

Resistors 3
Reverse biased 21
Reverse Biasing 16
Reverse resistance 27

Index

S

Saturation Level 82
Self-biasing of JFET 155
Semiconductor Diode 21
Semiconductor Materials 9
SI Units 7
Simplification in Sum of Product 234
Simplification of Boolean Expression using K-map 233
Slew Rate 198
Staircase-ramp DVMs 271
Subtractor 195
Sum of Products 231

T

Transistor Action 70
Transition and Diffusion Capacitance 27

U

Unity Gain 191

V

V-I Characteristics 22
Valence electrons 10
Voltage and Current Relationships 5
Voltage Divider Bias 86
Voltage gain A_v 93
Voltage-Divider Bias 82

W

Work, Power and Energy 6
Working of NPN Transistor 71
Working of PNP Transistor 71
Working Principle of JFET 149

X

XOR 228